Zero Defect Marketing
The Secrets of Selling High Tech Services

Zero Defect Marketing
The Secrets of Selling High Tech Services

Lee A. Friedman

with

David H. Rothman

DOW JONES-IRWIN
Homewood, Illinois
60430

*To Sandra Lynne, with love
and gratitude for her support*

This book was set in Century Schoolbook by TCSystems, Inc.
The editors were Susan Glinert Stevens PhD, Susan Trentacosti,
Joan A. Hopkins.
The production manager was Carma W. Fazio.
The designer was Mark Swimmer.
The drawings were done by Jill Smith.
R. R. Donnelley & Sons Company was the printer and binder.

ISBN 1-55623-001-X

Library of Congress Catalog Card No. 87–71543

Printed in the United States of America

1 2 3 4 5 6 7 8 9 0 DO 5 4 3 2 1 0 9 8

A NOTE TO READERS

Dear Reader:

People in the professional and engineering services industry need your help.

Here's a chance for some constructive bragging in order to educate others. For a follow-up book, I'd like to learn of your successful marketing campaigns or tactics. Tell the world what works—and what doesn't! Share your hard-won wisdom and illuminating anecdotes!

Other topics could include:

- Fiascoes. I've had mine. We all have. I'm especially interested in how Murphy's Law applies to marketing. If anything can go wrong . . .
- Experiences with good and bad clients.
- Ways you beat out competitors.
- Characteristics of competitors.
- Marketing myths and folklore.
- Management issues related to marketing.
- Whether this book's techniques helped.
- Anything else you wish to share.

Mail your letters—typewritten, please—to:

"Zero Defect" Marketing Program
c/o Techno-Marketing Concepts
7414 Georgia Ave., N.W.
Washington, D.C. 20012

For verification purposes, include your name, address, and phone number in your letter. And tell whether I should give you—and the subjects of your case histories—either credit or anonymity.

If we use your story, you'll get a copy of the book in which it appears.

Sincerely yours,
Lee A. Friedman

ACKNOWLEDGMENTS

THANKS TO . . .

- Our editor, Dr. Susan Glinert, herself a gifted marketer, who immediately grasped the need for a book like this.
- Dr. Lance Hoffman, Professor of Electrical Engineering and Computer Science at George Washington University. Like Susan, he provided painfully insightful advice.
- Friedman's valued mentor, Walt Davis, one of the top marketing men in high tech services, who taught him many lessons distilled here.
- Nancy Breckenridge of The Daisy Wheel of Silver Spring, Md., our transcriptionist. She gave us fast, expert service—even sending a long computer file to us over the phone lines during the winter's worst storm. (You think we'd use a low tech transcriptionist?) And she gave us inspiration, too.
- Sharon Hess, another superior transcriptionist.
- David Epstein of Dow Jones-Irwin, whose electronic manuscript advice helped streamline this book's production.
- XyQuest, Inc., Bedford, Mass., whose XyWrite word-processor rescued Friedman after Tandy's DeskMate II twice trashed one of his most important files—ironically, a section on the importance of responding to customer needs.
- Ourselves—for finishing this project as friends. High tech marketers and lay-level writers are like cobras and mongeese—inherently at odds. The marketers want to spell all the details out for prospective clients. Writers, on the

other hand, want to simplify for a broad readership. Our compromise was this: Friedman would provide his original lecture material containing all facts and opinions for this book (hence the single byline); Rothman would do most of the writing. Friedman had the last word, however. Since our book is a business and technical guide—not *The Great Gatsby*—accuracy and thoroughness have come before felicity of expression. We think the readers will prefer that priority. Above all, we intend *Zero Defect Marketing* to be valuable to the people who need it the most: the millions of Americans in high tech services.

Lee A. Friedman
and
David H. Rothman

CONTENTS

CHAPTER 1

THE ZERO DEFECT SNARK PLAYER

I had to fire Brad Jones the other day.

In his earlier job Brad sold millions of dollars of computers and accessories to the U.S. government as well as Fortune 500 corporations. He typically earned more than 80K a year.

Brad, however, couldn't cut it as a regional marketing man for my employer. We're a professional and engineering services firm that scrambles after one or two major contracts at a time. But Brad came from industrial sales. His forte was products, not *services*.

Yes, he could sell 3,000 modems to an Exxon man over drinks. But that wasn't the same as beating off 10 rivals to win a contract to set up a $15 million information system.

Brad wasn't dumb, mind you—he knew his circuits and switches and could tell good electronic equipment from bad.

"Zero defect," in fact, was one of his favorite phrases. We used to argue about it whenever he came to Washington. It was a techie myth from the 60s, no less out-of-date today than a slide rule or a punch card. You could never make zero defect a reality, just an attitude; look at the Russian nuclear mishap and the space shuttle tragedy. But Brad was a believer. "If you have a zero defect product," he said, "the rest will follow."

But as our marketing director in the Southeast he was full of glitches, an embarrassment to the old colonel who had supposedly hired a crack salesman.

He didn't go after new business. Perplexed, he just sat around in his $300 swivel chair, waiting for customer calls and reading *Commerce Business Daily,* the newspaper where the government tells what goods and services it wants to buy.

"Don't like to let you go," I said. "You fail, no one looks good. But you can't wing it like your boss. Harry, he has his old military contacts. Doesn't have to scrounge. But you—you have to play by the rules of the game."

Brad protested how great he was at arranging for strange computers to talk intimately to each other over satellites, how he was a technical whiz.

"But you couldn't talk to your customers," I said.

"First time anyone's said that about me," Brad snapped back.

"Well," I said, "you never bothered to find out their needs and rules. You didn't really find out who needed solutions."

"I followed up on 15 leads," Brad said.

"Mostly from *Commerce Business Daily,*" I said, "as if we're the only outfit in town that reads it. Look, it's just a start."

"I talked to contracts officers," he protested, "whenever I saw a prospect."

"And all they did was to put you on the bidders list," I said. "Thing is, you never really asked 'em what they wanted. You just called and driveled about what a great outfit we were. How can you say we're qualified to do a job if you don't even understand the problems they want solved?

"Remember that Defense deal?" I asked. "The communications system? Well, I talked to my buddy on the tech side and he says you didn't call him once. You just messed around with the business type. You didn't call my buddy and say, 'Look, here's how we'll make your system more reliable.' And then you go and talk to the wrong people."

"Well," Brad said, "how could I neglect the contracts officer?"

"And then you went off and told our competition who our teaming partner would be on a major bid."

"I was trying to scare them," Brad said.

"Well, I'd say you *comforted* them. You told them our technical solution and our pricing strategy and . . ."

"I've never . . ."

"Well, unwittingly you did. By telling them who we were teamed up with." Then I asked, "You play poker?"

He nodded.

"Well, we do, too," I said, "whenever we go after our contracts. And that's the whole secret." Brad and I never got around to filling out the poker metaphor—I can appreciate his priorities at the moment—but here's what I wish I could have told him in time to help.

WELCOME TO SNARK

"Look," I would have said, "our business is like playing a bizarre poker game where the bidders start playing at the worst odds.

"It's a one-hand game called Snark. A floating five-card stud game. But all the cards are kept face down. On each pass, players don't know how the other players bid.

"Each whispers his[1] bet to the house dealer and then places the bet into a paper bag. The dealer is the only guy who knows the bidding and the amount in the pot—the paper bag. He's the only one who sees all cards that have been dealt. And he sets the rules for the hand, for each deal.

"Another odd thing: The players are all wearing masks.

"You don't know who they are. And you can't notice any clues as to how good their cards are or how they are betting.

"After the fifth card is laid down, the dealer rips off one guy's mask and says, 'You, you're the winner.' And hands him the pot.

"Then all the players get up and move on—some to different tables with a new mix of players and a new house dealer and another game of Snark poker. With perhaps another set of rules and a different size pot.

"By the way, the house dealer will sometimes walk off with the pot himself and take everyone's cards with him.

"None of the players win in that case."

Seemingly, even the biggest losers in Vegas would stay away from a Snark table. Why buck such bad odds?

---

[1] Strictly for stylistic reasons, the examples in this book will normally use masculine pronouns like "he" rather than the more awkward "he or she." This isn't to suggest that women shouldn't venture into high tech services—quite the contrary!

Yet thousands of U.S. companies play Snark every day when they place their bets and lay down cards—offer their bids—to win contracts for high tech services in computers, defense contracting, architectural and engineering work, health services, and many other fields. Some 10 million Americans either are in the game themselves or work for employers who are.

Now suppose they could ferret out clues to unmask at least some of the other players. What if they could intelligently speculate on what their rivals' bid strategies were, and perhaps the sizes of the total pots? And suppose they could debug their marketing programs to eliminate—or at least reduce—mistakes like Jones's?

Obviously, then, they could play the game better. And *Zero Defect Marketing* will tell how. Believe me, bidding for contracts can be tricky as a Snark game.

WHOM THIS BOOK WILL HELP

Many otherwise sharp people—especially technical whizzes with Brad's zero defect approach—play Snark without even knowing it.

Or they may know the game is on, but not understand the rules.

So pay special attention to this book if:

1. *You or your bosses offer engineering, technical or professional services, or even management and business services.* Those Snark pots keep increasing in number and size. Our economy is indeed becoming more complex, oriented more toward high tech and services—full of opportunities. But more bidders show up at each Snark table. At least 15,000 companies are hustling (well, theoretically *hustling*) for federal contracts. And similar struggles are going on in the private sector.

Simultaneously, corporations such as GM, AT&T, and the Bell operating companies are jumping into the professional and engineering services fray. And that includes bidding on government contracts. So *Zero Defect Marketing* will include advice for people at larger companies who are new to the marketing of high tech professional services.

And if you're with a small company? Then you should learn to market skillfully to compete against both the giants and others. If you don't, you might suffer the fate of four fifths of the small firms in high tech services—and die within two years.

What's more, at both small companies and large ones, this book should be especially useful to women, few of whom now market high tech services. Don't blame me for the scarcity of female examples in these pages. Maybe, by alerting women to the opportunities here, I can help turn things around.

2. *You're a buyer of high tech services.* This book will give you no small clue into the mindset of those chasing your dollar, and besides, who's to say you'll be on the buying side forever? You may be able to begin a second career in retirement.

3. *You sell or buy high tech products.* You'll find advice on branching out to the services area. The payoffs can be huge. Suppose you make laser printers, disk drives, and other peripherals used in multimillion-dollar computers. If your own company designs the systems, your products stand a better chance of being included. The trick is to know how to be fair to the client and avoid what some would regard as a conflict of interest. Don't be bashful. None other than IBM is stepping up its services activities. And why not? Offering services can make you more sensitive to customers' needs and open doors to new business opportunities.

This book also works in the other direction, of course. My eleventh chapter tells how services firms can successfully develop and sell high tech products for government or business.

4. *You're a consultant.* This book will give you insights into how your consultancy can bring you larger and longer—and fewer ulcer-inducing—contracts.

5. *You're a business school student, especially one interested in becoming a manager or marketer in the professional or engineering services business.* Marketing your company's services won't be the same as selling soap or cars—a point that many professors fail to grasp. The standard wisdom from Procter and Gamble and GM doesn't always apply to the services industry. The more savvy professors understand this already.

6. *You're a street-smart B school professor.* Read this book to find out what's really happening in high tech services. Then add it to your course syllabus.

7. *You're an engineering school professor.* Your students should know what their future employers will require of them.

8. *You're curious about the way high tech people chase after your tax dollars.* Or perhaps you want to know about the equivalent fun and games in the private sector.

9. *You're an innocent bystander, who, though far removed from the marketing battles, can appreciate my war stories as sociology, anthropology, or psychology.* Academics study old Sears catalogues to find out how Americans lived in the early 1900s. And perhaps years from now, professors will read *Zero Defect Marketing* to learn the strange rites and rituals of the services industry as viewed by an insider.

All my stories are true in principle if not in detail. Of course I've created some composite case histories to spare my friends' feelings and myself some lawsuits.[2] And in the interest of clarity, I have also followed the examples of many writers in this genre and created dialogue. Still, you've picked up the right book to learn how my industry really works. I can tell. I've been outfoxing rival bidders for 25 years in the United States and overseas, pursuing both commercial and government contracts, working for small and large high tech companies, supervising hundreds of people in some jobs, serving as a general officer in the high tech marketing wars, close enough to the front lines to see the problems first hand but wearing enough stars to enjoy strategic perspective. I have:

- Managed the design and development of automated systems worth billions of dollars. My systems keep track of everything from earth satellites, to government personnel records, to federal budgets, to inventories and supplies, to scientific and engineering information. Not just budgets but lives may depend on how well I did my work.

---

[2] I intend all corporate names in this book to be fictitious except for (1) those of Fortune 500 companies, (2) a Maryland firm called Marketing Consultants International, (3) information sources for readers, and (4) my reference to Tymnet. Any other resemblance between case-history inventions and real names is strictly coincidental. That applies to "Lancelot Technology" as well as "Slee Z Systems."

- Been a high tech marketing director and successfully broken into new markets for professional and engineering services.
- Designed and marketed high tech products for public utilities, chemical and steel companies, and tax preparation firms. My brainchildren were more profitable than company and industry averages. And their rivals ranged from mom-and-pop consulting firms to Fortune 500 giants.
- Directed and been a member of corporate steering committees that dreamed up new commercial and federal markets for services firms.
- Worked as a marketer in health services.

Don't let the word *marketer* put you off. Yes, this book contains lessons for marketing professionals. Especially, however, I'm writing for technical people—the ones who so often lose out on promotions because they aren't attuned to the importance of marketing.[3]

No one can ignore it! Even the most prestigious scientists must market ideas. Without marketing—conscious or unconscious—the theory of relativity might be just the musings of an obscure physicist.

But this book isn't about marketing ideas or about how to move more soap or fertilizer. Rather, it's for people and companies offering high tech *services*.

What Are "Professional and Engineering Services?"

This book's subtitle is "The Secrets of Selling High Tech Services," and it indeed will help you no matter what kind of high tech service you or your company offers.

But I'm particularly writing for suppliers of professional and engineering services. They would:

1. Apply the principles and standards of professional disci-

---

[3] Particularly, marketing's importance in creating technical solutions.

plines such as engineering and mathematics and the natural, social, and physical sciences.

2. Do this in the design, development, evaluation, and management of complex processes. The word *processes* here would include everything from building a tank to supervising a production line; from running a record-keeping system, to diagnosing a disease; from issuing purchase orders to storing supplies in a warehouse.

In short, I'll tell what they didn't teach you in engineering school. Or medical school—or wherever else "experts" are too professional to pass on street smarts. I won't tell how to cheat. Just how to play your fairest, toughest game. I'll even tell how, without breaking the law, you can spy on competitors.

WHY TOUGH'S BEST

Daily, a nasty truth keeps burrowing deeper and deeper into my brain: Second place is no good; either you win a contract or you don't. When you're second consistently, you'll go out of business. You aren't like a car dealer who can limp along selling fewer Cadillacs than his competitor. Even if you boilerplate together a proposal, you're offering a one-of-a-kind solution for a specific market with changing needs to which you must quickly respond with the latest technology. So what if the customer by some fluke is nice enough to praise the care with which you put together your second-place proposal? The point is, you lost.

And if you lose too many times? It's your job or, if you're an owner, the end of your business. I know of divorces, children deprived of college tuition, postponed retirements, and other aftermaths of lost contracts—many of which might have been avoided if high tech companies had used my Zero Defect approach. You need it because you don't even have a chance for repeat sales. You'd better be right the first time.

Especially in high tech *services,* however, nothing is sure-fire. Your client can legally and even "ethically" change the buying rules behind your back. Watch out. Some ambitious bureaucrats regularly try to outwit contractors in hopes of moving up the ladder. They think private industry owes them freebie after freebie. Not true. Among major government contracting services firms in and near Washington, D.C., the yearly net profits are between 1 and 3 percent of annual revenue.

Simply put, you're more likely to make wrong choices in services contracting than when offering bids for lug nuts. Your customers are often passing judgment on intangibles. And that opens the way for slipshod decisions by you or them.

But can you win anyhow? Maybe. I'll explain later. Meanwhile, let's just say the ideal marketer of high tech services would be a cross between a top military strategist, a crack CIA agent, a hungry lion, and the best card shark in Vegas.

A MARKETER ISN'T A SALESMAN !

Notice I said "ideal *marketer*." We're talking about marketing, not selling, a distinction that eludes too many high tech companies.

You can't just be an order taker like the old drummers. No one has successfully peddled computer services to Ford Motor Company by slapping backs and asking, "Hey, Henry, you got any $8 million high tech contracts today?"

As a true marketer, you must respond more precisely to the buyer's needs, individual as well as organizational. A gap may exist between the two. Fortunately, however, organizational and individual needs can coincide. Suppose you sell a good computer system design to a young Army major and time proves you both right. Then you'll increase his chances of making general.

BEATING THE DISMAL PERCENTAGES

In this book I'll do my best to respond to your own needs. There are many of you out there: smart, hardworking people who want to win contracts but who fear they lack contacts. I'll tell you how to outwit your rivals.

And you'll also learn how to convince the buyer (the Snark

dealer) that you've got the best cards (the technical solution or support services) and price (your cost proposal). How, then, do you beat the odds?

Your odds aren't quite as bad as those at the Vegas slot machines—your win rate may be as good as 35 percent[4]—but then again, a bid for high tech services may set you back the equivalent of an Everest-sized mountain of poker chips. A company such as GM's Electronic Data Systems may spend tens of millions of dollars on just one bid.

While EDS and Boeing are fighting for billion-dollar contracts—and with the outcomes influencing stock prices on Wall Street—thousands of small-time firms are chasing after paltry $100,000 contracts. Forget the stereotype of a high tech firm housed in a block-long building or on floor after floor of a glassy skyscraper. There are more small-fry than you'd think. In software services and products, for instance, some 40 percent of the companies have fewer than 10 employees. Another 33 percent have between 11 and 30 employees. The small-fry player may lack even anthill-sized piles of chips, but they need every chip just to stay in business. What's more, they don't want forever to be on the treadmill of chasing after small contracts or the less profitable leftovers that the big companies shun.

An intelligent marketing program can help. In fact, in some cases, it may even let the small-fry grow larger without needing outside venture capital.

Small company or large, the challenge is to pursue the right clients at the least cost. Breaking into a new services area—finding clients, learning how to prepare proposals, and so on—may gobble up between 20 and 45 cents of every revenue dollar for marketing.

Proposal presentation and demonstrations alone, encompassing just odds and ends such as slide shows, can cost 2 percent of the sales value of a small contract. For a larger contract, the percentage is less. But the dollar value goes up.

Seeking a $4 billion communications engineering contract,

---

[4] The win rate might actually reach 65 percent when you are bidding on follow-on, competitive solicitations. See Chapter 7 for more on that kind.

a Fortune 100 company invested over $12 million to demonstrate the feasibility of a digital and voice network even before the government officially asked for bids. You, of course, can try to slash such outlays. But why not make certain that the money is going for a good, attainable goal—one of the many benefits that Zero Defect Marketing can bring?

Whether in private industry or government, procurement officers do not always show much interest in the preservation of bidders' chip piles. Take the case of a large Eastern power company.

> The power company asked an engineering firm to submit a detailed proposal to build an electronic security system to help keep out trespassers. Confidently, the engineers went about spending $91,000 to work out the proposal. The fish looked fat, the barrel thin. This was to be a sole-source contract, and the power company even offered budgetary estimates.
>
> A month after the engineers put in the proposal, a man called from the power company.
>
> "Don," he told the president of the engineering firm, "I want to congratulate you on your proposal. Very thorough work."
>
> "Thanks."
>
> "And I liked the way you put the foot down to make our deadline."
>
> "We do our best."
>
> "That's why I'm so embarrassed."
>
> "Embarrassed?" asked the engineering company's president.
>
> "You see, Don, our people have put this on hold. Just don't know when I can get back to you."
>
> Talking to the power company, Don kept his temper in check, but afterward he felt almost like throwing the phone across the room.
>
> It didn't help three months later when Don ran into a supplier of power plant equipment—and learned his supposed client had used his proposal and specs to do the work in-house. Don felt betrayed. The power company had wanted proof that his firm could make good cookies. He'd come up with a recipe, only to see the job go to another baker.

Don is just one of many people in dire need of a zero defect marketing strategy, which, though hardly insurance against fiascos like this, can reduce the risks. He might not have bid on the proposal if he'd gathered enough intelligence on the power company to know its true intentions.

Again, no system's foolproof. But what if Don had been following the zero defect marketing philosophy? Then he might have checked carefully with a trade association or suppliers to find out if the power company had a track record of mooching freebies.

If you don't know the "dealer's" mind, of course, you may also lose your chips to one of the masked people at the green table—a rival bidder.

So don't be too smug about that ace. Your rival's hand may be yet stronger.

THE HIDDEN COMPETITION

Remember, until the buyer announces the winner, you probably won't even know who the competitors were. And you will not know if they're better or worse connected than you.

A buddy of mine at Wired Software looked forward to a fat contract with the Marine Corps to build a multimillion-dollar system to keep track of supplies at a large military base.

"I didn't see how we could lose out," he said. "I mean, we knew the managers. Everyone. The purchasing people. The logistics people. The people who'd use the system. They even told us the budget on the Q.T.—$5 mil.

"There were six companies after this baby," my friend said, "and we thought we could outsmart 'em. We'd play safe. We'd bid just $4 mil. It was a cost-reimbursable deal, the usual. We could make up our costs later. We had a winner. Good work and a good bid price, and we lost."

"With a 'winner'? " I asked. "What happened?"

A few not-so-good men had sandbagged my buddy. The actual budget was "two and a half mil," he said, "not five. Jeez, they didn't even get to the technical evaluation. Just went by cost. And just one outfit under budget—ACP corporation."

"How much did they bid?"

"2.4 mil." My buddy sighed. "Let's just say we didn't know where the real wires ran."

"So you were there for the cosmetics?"

"Right. You turn the auditors loose on this one, they'll think the game was wide open."

In short, Wired was miswired—to "friends" who weren't friends, and who carefully lured in sucker-bidders to pretty up the proceedings in case the General Accounting Office or others questioned the ACP award.

What if Wired Software had been able to unmask the poker players and anticipate the dealer's moves? And can you avoid similar losses? No promises here. But later in this book I'll show you in detail some possible zero defect strategies along with smart battle tactics to help you beat out hidden competitors.

The Lost Tribe

Here's another example of the perils of fighting hidden competitors:

> Several years back, an American Indian firm also might have profited from the lessons of this book. The U.S. Department of Energy encouraged the company—we'll call it Silicon Tepees, Inc., without any slurs intended—to bid on a scientific and engineering information system to aid oil company technicians and others.
>
> The system, for example, might store information on what kind of diamond drill bit worked best on a certain type of rock. Just as with Don's engineering firm, the lights seemingly shone a bright green.
>
> A friendly Energy official unofficially revealed the government's budget for the contract, $500,000, and Silicon Tepees even knew who its competitors were.
>
> Teaming up with another small firm well respected at Energy, Silicon delivered on time a three-volume proposal—superior work.
>
> One Friday six weeks later, the head of the proposal evaluation committee dialed up Silicon. "This is out of school," he said, "but you're it. You're the best technically. And you're the lowest." Silicon had bid $415,000. "Midweek you'll hear from us, OK?"
>
> Well, word came the next Tuesday—that rival company had won with a $175,000 bid.
>
> What had happened? Silicon, alas, hadn't known what was happening in the Department of Energy beyond the evaluation committee. A new director was about to take over the Department of Energy lab in charge of the bids.
>
> Even before his name went on the door, he had learned how

much Silicon and rivals had each proposed. And that Monday, the first business day after his arrival at the lab, he pulled out of his hat the $175,000 bid from a stranger to the proceedings, Greenhorn Systems.

Silicon Tepees reacted as if the feds had proclaimed George Custer to be the victor at Little Big Horn.

"Unfair!" protested a congressman and senator from Silicon's state—but the new director's decision stood. And the political hubbub didn't exactly increase Silicon's chances of winning future Energy contracts.

There is, incidentally, a nasty little footnote.

Within a year Energy said the information system design was a botch and needed hundreds of thousands of dollars of additional work—which, on top of the $175,000, meant that the taxpayers were spending more than if Silicon had won.

Had Silicon followed my Zero Defect strategy, it would have more closely monitored the personnel shuffles at the Department of Energy.

Granted, it might not have prevented the new man from doling out the $175,000 contract to Greenhorn. But by befriending him early enough, Silicon might have earned enough brownie points to be ready to step in and pick up the pieces after the rival company botched the job.

Can you *really* unmask the players and profitably befriend the dealer, however? Well, take heart.

I can't promise to turn you into a tech services tycoon overnight. But zero defect marketing might well allow you to jack up your success rate from, say, 20 percent to 35 and maybe 50 percent—enough to add millions of dollars over the years to the coffers of a small- or medium-sized company.

THE ZERO DEFECT SNARK PLAYER (THE REAL ONE)

Here's how zero defect marketing can work for you. Let's say you're a marketing vice president of an architectural/engineering company, and you're bidding for the business of Hypothetical Automobile Parts or HAP.

Well-off teenagers have given up on TV, movies, and computers. With a new vengeance they're tinkering with automobiles, the expensive, turbocharged kind. And HAP's business is zooming along like a hot rod with the police in pursuit.

Now HAP needs a giant warehouse complex to store all the radar detectors and oversized tires for which the spoiled brats are pining.

The architectural/engineering work is worth a good $2 million. Normally your marketers can line up at the plum tree well before the competition does; but through bad luck—again, nothing is certain even in zero defect marketing—you don't find out about the job until the last minute.

You learn from a friend at a trade association conference that your top rival is the Slide and Rule, an engineering company four times your size.

S&R is the IBM of its niche. Customers often come to it rather than the other way around. But S&R isn't exactly comatose here. It's in close touch with a regional HAP vice president for whom the engineers undertook an earlier warehouse project.

Other wheels are grinding. Already S&R has drafted tentative agreements with prospective suppliers and subcontractors for the warehouse project.

But you're convinced you can still outsmart S&R. It's 100% engineering services. You, on the other hand, mix engineering with other professional and technical services and you're more used to scrambling for business than your stodgy, elephantine rival is.

That still leaves the question of how you assemble your own list of subcontractors and suppliers to prove to HAP that you can put together the warehouse.

Unwittingly, HAP itself rescues you. At the last minute you show up at a bid conference where HAP will distribute bid specs, and lo and behold, you're able to copy down the list of people attending—including those from "prequalified" companies that HAP had gone to the trouble of scouting and informally prequalifying as suppliers or subcontractors.

Presto! You know who can do the iron work, provide the computers, and offer a number of goods and services that you need to put together your warehouse proposal.

Within days you choose the suppliers and others you'll use. Then, so he isn't surprised, you run their names past the regional vice president who is *now* responsible for the warehouse bidding—not the same man who'd given business to S&R.

Via your trade association contacts and long chats with the V.P., you learn the man's likes, dislikes, motivations, hobbies, interests. He is, for one thing, an avid dove and grouse hunter. An engineer under you is himself keen on upland game birds, and you make sure he receives a crash course in everything you know about HAP.

He accompanies you to your next lunch with the HAP buyer. The two men exhange hunting stories like old friends.

Your staffer isn't faking a thing. He and the HAP man establish an easy rapport, and soon inside information is pouring out about your prospective client—HAP's purchasing process, budgeting practices, who makes what decisions, what problems the warehouse complex is to solve.

And you learn more, too, about the HAP vice president's personal motives. He'd like to speed up deliveries to wholesalers, but you can also pick up hints that, being on the verge of retirement, the V.P. wouldn't mind leaving a personal monument behind. The warehouse project was his idea from the start.

The engineer is only the beginning. Soon your best accountant befriends HAP accountants, your computer staffers wine and dine theirs, and on and on it goes—your people working smoothly as a team to influence their peers at the auto supply company and learn their needs.

Meanwhile you're all studying the visitor's log to see which of your rivals are visiting HAP, and how often, and to whom they talked. Only two companies other than S&R show up in the log. But you know from the past what their strengths and weaknesses are and can draft your proposal accordingly. You're far ahead of rivals—nine altogether—when HAP finally sends out specs.

Because you've listened harder and longer, your proposal matches the specs more closely than anyone else's. You haven't tried to adjust HAP's problem to your solution. No, you've been a true marketer and done the reverse—expertly tailoring your staffing and subcontractor and supplier plans to HAP's requirements. You know you can't sell Bibles to preachers if your sales brochure says you don't believe in hell. And what's a proposal but a brochure in another form?

Your main rival, S&R, satisfies the bid requirements, but not as closely and in as much detail as you did. So you win the contract with a price $139,000 higher than S&R's.

You haven't broken the law. You haven't done anything that most Fortune 500 corporations would condemn. You've played by the rules, unmasked many of the players, befriended the dealer and won the pot. Congratulations! You're a true zero defect marketer.

MARKETING REDUX

An inspirational story, of course, isn't enough. *Psst!* Here's the rest of my plan to help keep you in the chips and ahead of your competitors.

In Chapter 2, "The Friedman Formula for Competitiveness," you'll learn how a marketer can be aggressive without being a sleazy huckster.

"On Track with the Marketing Engine," Chapter 3, explains the basics of the marketing process, from finding a hot prospect to closing the deal. You'll learn what to say—step by step—to your prospective client.

Chapter 4 tells how to set up and profit off "Your Market Intelligence Database." Don't let competitors or new laws startle you, or at least not more than necessary (nothing is sure fire in high tech services). And be on the lookout not just for threats but also for opportunities.

In Chapter 5, "Competitors," you'll learn how to deal with different kinds of adversaries—every type from the Decommissioned Garrison (consisting of ex-employees of the organization you're trying to market to) to Superpowers (giants such as IBM).

Chapter 6 is "Buyers." It tells how to deal with the strategies of different kinds of clients. The buyer isn't always your friend. In some ways he's an adversary constantly trying to outsmart you; buyers do not, and often cannot, show "brand loyalty."

In the next chapter, 7, "To Bid or Not to Bid?" you'll learn how to estimate your chances of winning a contract and to act accordingly. At the beginning you'll hear about "Lee's Law." What is it? Simple. "If anything can deprive you of a contract, it will."

Chapter 8 is about "Nine Marketing Myths and How to Overcome Them." For instance, don't think everyone will bid on a job because *Commerce Business Daily* has advertised it. Your competitors may be avoiding it for that very reason. (Brad Jones was not foolish to read the *Daily,* just to depend excessively on it for leads.)

Chapter 9 tells of "Fifteen Ways to Outwit Rivals." Sometimes you'll actually come out ahead in the long run by giving away freebies to buyers; this chapter tells when and how to use this and other special tactics. You'll also learn how to promote your company to give you an edge in the long run.

Chapter 10, "The 100-Pound Sales Brochure: How to Develop a Winning Proposal," warns against common errors and tells in detail what a good proposal should contain. Writing a major proposal without a design plan is like trying to build the World Trade Center without blueprints. Even minor ones require considerable thought ahead of time.

In Chapter 11, "Products: Mud Pies for Sale," you'll find out how you can successfully develop new products—no small feat for high tech services companies, even the giants, most of which fail when they try to break into products. So often such firms design for themselves rather than for the market. Read this chapter to avoid the syndrome.

But first, why not learn what you can do to market your *services* competitively—the topic of the next chapter?

CHAPTER 2

THE FRIEDMAN FORMULA FOR COMPETITIVENESS

Philistine Technology once towered above its rivals. Ensconced on a bluff overlooking the Potomac and full of ex-Pentagon men and the usual techies, Philistine enjoyed $40 million of new business each year. By 1986, however, Philistine was living off past successes. Earnings had leveled off. Only because of inflation—and the extensions of Philistine's current contracts—did they not go down. Amid unflattering newspaper headlines, the owners sold out.

They had violated Rule #1 of the Friedman Formula for Competitiveness. Here are the Formula's parts:

1. Try harder. Follow up every opportunity for new business, and be persistent about this pursuit! Competitiveness isn't just for second-place car rental agencies.
2. Micromanage your people. Govern your people's actions and decisions. For instance, you should tell them in detail how to brief clients and how to write proposals.
3. Nullify the potential for failure. Convince the client that neither he or nor you will go belly up.
4. Sharpen your competitive edge. Be able to show you can offer important advantages, in price or other areas.
5. Be moral.
6. View competition as a corporate reckoner. Evaluate your company and people by how successfully they deal with the rest of the world, especially buyers and rivals.

The Philistine leaders were competing for contracts without acting competitively. Puffing away on $3 cigars, surrounded by exotic plants and toothsome secretaries, and favored with corner offices, they had viewed the world all too smugly through their giant picture windows. So did people lower down. They didn't care to try to carve out new markets amid increasing competition. Rather they were content to expand existing contracts and to see competition as a duty rather than a natural instinct and a corporate reckoner.

Why should engineers, programmers and other professionals shame themselves by seeking out clients, listening to their problems and needs, and offering some workable solutions? The nerve of such heretics! Who do they think they are? Willy Loman? Salesmen?

By applying the formula—and later I'll give a success story—you can help avoid Philistine-style mistakes.

THE BASICS OF THE FRIEDMAN FORMULA

The better companies will infuse everyone with an eagerness to pounce on opportunities. And that includes not just marketers, executives and senior managers, but also the staff—the frontline troops, especially the so-called techies. All too often, services companies share the same basic management policies, pricing, hiring and even firing practices. And ideally you'll come up with better approaches in *all* those areas in your tactics.

One way to do this is to follow the five rules of the Friedman Formula:

Rule #1: Try Harder (It Didn't Hurt Avis)

"Trying Harder" isn't just for Hertz, Avis,[1] and football coaches.

Imitate lions[2] and constantly search for new prey. Don't shrink away from hard chases, as long as the antelope looks

---

[1] Remember Avis's old rent-a-car button—"We Try Harder?"

[2] I use poetic license whenever "he" refers to a lion searching for prey. In a pride, the female normally is the hunter. Also, however, I'm aware of the traditional use of the masculine in hypothetical examples. No sexism intended!

juicy enough. And run at full speed, each time. Ignore the psychobabblers who claim that competitiveness leads to personality disorders. Those laid-back wimps, not you, are the true sickies, and if Americans ever end up as janitors, sweeping up behind robots in Japanese-built factories, the babblers will share no small part of the blame. It needn't happen, though—not if enough people follow the lessons in this book.

A Few Lionly Pointers on the Art of Trying

I haven't heard any lions whispering any marketing tips to me lately. But they might as well be. Just watch a documentary on public television, and you'll wish a lion were on your side next time you bid on a $20 million information system.

If nothing else, the lion is persistent.

Coming across the hunted, he'll catch the prey only one of five times.[3] He isn't a crybaby. Imagine a lion calling up a congressman to complain that he didn't catch an antelope.[4]

No, the lion knows he's competing for limited resources.

You aren't about to sink your teeth, lionlike, into the bidder yourself—but you, too, are after limited resources. There are only so many antelopes running around in your part of the bush. And some other animals probably are also in on the chase.

Like the lion, too, you must:

1. *Keep your eyes open.* Read the "bids and proposals" section of the classifieds in your daily newspaper, *Commerce Business Daily,* or equivalent publications in your own field. Find out who's regularly issuing requests for bids or proposals and for what types of services. Can you befriend the buyers and learn their needs *before* the ads reach print? Also, learn who won bids, and at what price. What better way to find out how competitors are doing, what kinds of jobs they're seeking, and who you may be up against if the buyer rebids the job?

2. *Listen!* Unwax those ears. Listen for the antelope rustling through the bush—that is, perk up your ears for clues that

---

[3] No sexism intended with the "he'll." Let it be known that she-lions show similarly admirable traits.

[4] Contrary to widespread belief—perhaps engendered by the line "Where the deer and the antelope roam"—antelopes are found in Africa as well as the United States.

a solicitation may open up. You can pick up clues anywhere. I've picked up clues at churches, fraternal organizations, alumni meetings, cocktail and dinner parties, and even on airplanes. And once you're talking to a buyer, pay attention to the nuances and inflections in your prospective client's words as he tells you what his needs are.

3. *Don't roar at the wrong times.* Your roar is your sales pitch; be sparing in letting the world know how great you are or how much similar work you've done for other clients. Better to spend most of the time discussing your client's own problems.

4. *Have a good sense of how far you can go.* The lion knows how to gain territory both aggressively and passively, and how to hold it. That's not such a bad skill when deciding how far you yourself can go in:

- Protecting yourself against other predators—those rival bidders.
- Proposing terms for the bidder or negotiating with him or her.
- Deviating from the exact letter of the contract.

Rule #2: Micromanage Your People for Full Competitiveness

Micromanagement will help you fight Murphy's Law and reach your goals. Micromanagement is a step-by-step management approach. With it, you maximize control over organizational processes to assure a specific result or outcome to reach an objective.

Granted, you should encourage your people to act independently—but use micromanagement to monitor them and to teach them the ropes; continue until they're fully acting out their competitive instincts. That is, you'll attend to the details of:

1. *Organizational responses.* To pursue a client you need the right people with the right talent at the right time, with the right support and resources, orchestrated the right way. A single mistake could doom the whole effort. So use micromanagement to keep on top of such matters as:

- Gathering intelligence on buyers and competitors. Ask

your people, "Who did you see today, and what did you find out?" And, "What do you need to win the job?"

- Selecting the right supplies and services. Will your suppliers—of products or people—be reliable and meet *your* specs?
- Internal support services, such as library resources and financial and contract management.
- Planning strategies and tactics to deal with short- and long-range threats and opportunities.
- Selecting the right people for proposal writing and other tasks. Be future minded in this detail work. For instance, look for proposal writers who know how to brief superiors, and who might therefore be candidates for management jobs.

All too often people in this industry lack the competitive spirit, so that you must micromanage them to compensate.

2. *Exactly how staffers should write a proposal—and especially what they should or shouldn't say.* Standardize proposal procedures when possible. And remember: The goal isn't to write either the Great American Novel or a world-recognized technical treatise. Rather it's to assemble facts that tell why your bid should prevail. Don't hide the highlights. Make them conspicuous in your letter of transmittal and in your executive summary. Don't organize a proposal as if it's a paper describing a college science project, with the conclusion (the benefits) just at the end.

In proposals and related correspondence alike, your people should write good, direct English and depend on skillfully assembled facts, not turgid rhetoric. Say, "Avoid the passive tense if you can." It's usually for wimps. Never forget that good English can give you a jump on the competition. If your clients can't understand your proposals, no one will buy.

For more on proposals—including what to say and what not to say—see Chapter 10.

3. *How your staffers should dress.* Let your in-house programmers wear beach clothing behind the scenes if they'll be more productive. But insist on a conservative, buttoned-down image for project managers and others in contacts with clients. The starched white shirt was no small part of IBM's success.

Ross Perot's company, Electronic Data Systems, similarly grew because its employees even *dressed* competitively—not for the fashion pages but for stodgy clients.

Rule #3: Nullify the Potential for Failure

Nullify the potential for failure, both in reality and in your client's mind.

Allow for the unexpected calamity—for instance, a Xerox machine breaking down when you're cranking out a proposal under deadline, the sudden loss of key staffers, or a troublesome subcontractor. And use the micromanagement techniques discussed earlier.

Above all, let your client know of your interest in reliability. Do you read *Consumer Reports* when you shop for a new car or refrigerator? If so, you probably don't just look for products with the latest gizmos. You're also shopping for reliable ones. You don't want breakdowns on the way to work.

And most clients and prospective clients feel the same when they compare bids and bidders.

The prospective client must pare down a bidder's list to one company. And why not do this by sniffing out anything that could suggest failure?

It's a very human reaction. After all, if you fail, the contracting officer or purchasing agent will fail too—as the one who chose you.

So as your client reads your proposal, he's looking for evidence of potential project failure, in areas like:

- Budget or cost overruns. He'll search for statements such as that within a year you'll have to replace equipment you've proposed.
- Excessive staff turnover. Your client might look askance at your promise to maintain a steady recruiting campaign during the contract. He might worry that you can't keep your people.
- Ineffective solutions that don't meet his specs.
- Sloppy project risk assessments that don't warn of impending problems.

- Dishonest claims. Don't lie about your company's ability to perform and produce; clients may check references.
- Unsubstantiated assertions—for instance, statements such as, "We can do the work," without telling how.
- Surprises! Clients hate surprises in the above categories or others. Most clients will spot the source of a potential surprise, and back off.

They'll worry, too, about another problem—your failure to win more bids. The clichés are true. Failure breeds failure; success breeds success. In fact, if your bid succeeds against your rivals' bid, you'll enjoy a bonus: another failure on their records. No meanness intended. This is simply one of the laws of competition—and another justification for zero defect marketing.

Rule #4: Sharpen that Competitive Edge

As a competitor, you should lead in more than one of three ways:

- Pricing gambits. Competitors' prices may not vary as much in professional and technical services as in other industries, but even here you can stand apart.
- Improved communications with clients, not just in proposals but in personal contacts.
- Quality.

Pricing Gambits
I can just hear the whines when I start talking about winning the price wars.

"Services aren't like products," the skeptical will object. "You can win a volume discount on steel to make cars, but not on salaries. The 100th engineer you hire isn't going to work any more cheaply than the first."

That's true. Prices do not vary nearly as much within the services industry as they do elsewhere.

Still, there are times when you can claim more of an advantage than you might think. Consider:

- Labor prices.
- Prices in proportion to the size of your company

- What will be a winning price as determined by the client's budget and what you expect competitors to bid.
- Prices in relation to company profit goals.

See Chapter 7 for more on pricing.

Improved Communications with Clients

Good communications can be as much—if not more—of an edge than the latest computer system. The better you are as a listener, the better you can meet your client's true needs. You can't compete unless you can communicate in letters, in proposals, in client meetings, in phone conversations.

You must show you're better than rivals, while nullifying the actual and perceived potential for failure.

If you're a manager, be prepared to use the micromanagement techniques discussed earlier.

Enough communicated?

Quality

Your products, services, and management should all meet industry standards. Build in quality at the start and as you go along. Otherwise you'll constantly have to fix up earlier mistakes—usually at your own cost. And that's if you're lucky. You might lose the contract altogether.

So don't wait for failures to reveal lack of quality.

Rule #5: Be Moral

Morality starts at the employee level.

Treat your people right. Expect employees to treat others right. Clients hate surprises, and one of the best ways to reduce the number of them is to hire honest, dependable employees, who, of course, being the way they are, will expect moral treatment by the boss.

In dealing with clients, never lie or be unethical; and tell your people not to, because sooner or later the facts will catch up. This isn't just morality—it's also practicality. If you hype up your qualifications or imply that an unworkable solution will work, you're doomed from the start.

And if you don't tell your clients of problems you're having on a contract, you've opened the door to further woes—and loss of future business.

If you draw hasty conclusions, for instance, indicating that some method will absolutely work without a true test, and if you know that's false, then you deserve a hostile client.

Similarly, be moral toward your competitors. You never know with which companies you'll team up on major projects in the future. Mind you, I'll qualify that. There are certain practices within the services industry that by normal standards skirt the line—for instance, showing up at the other guy's company for a job interview to snoop on the competition. But that's part of the game: Play, or become a social worker. You're really no different from a reporter posing as a jail inmate, or from a policeman going undercover.

Still, there are limits. Avoid bad-mouthing competitors unfairly. Don't tell the buyer that your rival is unprepared to do the work, unless you can prove it. Besides, your prospective client may think that if you bad-mouth your rivals, you'll bad-mouth him.

Rule #6: Competition as the Corporate Reckoner

Mayor Koch of New York has a favorite question that he pops to voters: "How am I doin'?"

Service contractors, however, needn't wait for an election to find out the best answer to such a question. The better ones are constantly grinding out proposals and receiving feedback—acceptance or rejection. And therein lies an opportunity. Competition is a wonderful, meaningful gauge of good management and a way to weed out the inept. Don't ignore this tool! If you lose out on a contract, learn why the other guy won. If you're losing many, then maybe you'd better reassess your entire way of doing business. This isn't for the fainthearted. Maybe that's why so many people in professional and technical services either ignore this truth or live by it only grudgingly.

They are zoo-bred lions. The territoriality still may be there, but not as strong as in bush animals. Zoos such as engineering schools fail to nurture sound, competitive instincts. Forget about grades themselves. Racking up a 3.9 average in advanced calculus isn't the kind of competition I'm talking about.

No, I mean other numbers: those on corporate ledgers, and in counts of contracts won, and increases in income and in market shares. The tragedy is that within high tech services, corporate cultures often favor noncompetitive people who consider themselves above the bottom line. Too many techies can't understand that the amiable professional colleagues they see at conferences and seminars are threats to their livelihoods; and so these zoo-bred lions reveal weaknesses that competitors can pass on to buyers; or they blab about new techniques that rival companies then pick up and use against them.

How to make your people behave like bush lions? Simple. Gauge your company and its people by success or failure, both in the short and long runs. After every bidding session, and at least several times a year, you should assess:

- Information gathering. Do you get enough information on competitors?
- Client relationship and degree of acceptance. Does your client see you as just another services provider, or as a leader in your field offering superior solutions?
- Organizational responses. Have you got the right kind of team and have you organized these people well to respond to bids?
- The extent to which your solutions meet your buyer's standards. Have you met *all* requirements? Then you've probably done better than the competition.
- Risk reduction. Have you done a better job than rival companies at persuading the buyer that you can reduce the possibility of failure?
- The extent to which your price coincides with the value of your offering. How do you compare with rivals?

APPLYING THE FRIEDMAN FORMULA

My friend Marty doesn't need to follow the Formula—he's one of the people I had in mind when I invented it. Marty instinctively knows how to market. If he were a lion, he'd be leader of the pack—or at least the best antelope catcher.

Here's how he caught a $10 million antelope, a services contract with a multinational firm.

As marketing director of an engineering services company, Marty didn't stay behind his desk waiting for customers to dial him up. He roamed. And it wasn't just aimless gallivanting. He showed up regularly enough in the offices of clients so that when they had a new need, he would be among the first outsiders to know.

"I have a problem," said one of his clients out of the blue. He was the manager in charge of the Mailing and Shipping Department at Generic International. "The problem is . . ."

Marty forgot the original reason for his visit and listened—waiting, so to speak, to pounce. It was as if someone had waved red meat in front of a lion.

The manager, Bob Gordon, went on.

"We're automating shipping," he said, "and need to account for shipments of critical letters. Or orders. Or packages."

"What," Marty gently teased, "you don't trust the U.S. mail?"

"It isn't just the U.S. mail," said Gordon. "Also the French mail. The British mail. The Kenyan mail. Our stuff goes everywhere. Now, I know there are certain things we can automate fine. The weighing. The sorting. The material handling subsystem within the main office. But how the devil do you make sure everything really gets there? How do you guard against loss or misplacement?"

"I know a good courier service," Marty said. "We'll just send them out with your packages handcuffed to the deliverymen. Seriously, here's the answer to your problem." And then Marty gave it—all the time showing a powerful ability to make rapid associations that rivals never would have dredged up. "The answer," he said, "is as near as your grocery store. You'll just put bar codes on your packages and envelopes. And then you'll use wands to decipher the lines. Just like the ones that scan cereal boxes and tomato cans to see what the prices are."

"What's that got to do with my packages?" asked Gordon.

"Here's what. Your headquarters people will key in the address of the recipient and produce a bar code label to be pasted on the envelope. As they do so, they'll set some wheels in motion. Via computers they'll automatically create records showing who's sending the mail. And who's to receive it. And there won't just be bar codes on the envelopes and packages. Also on the packs in which they're sent."

"What happens at the other end?"

"The receiving clerk in Kenya, France, wherever, will use the bar code wands to record the arrival of the mail packs. And he'll do the same to the letters inside—when he delivers them to their recipients."

"But how will headquarters know everything arrived safely?"

"Because when the receiving clerk completes his deliveries," said Marty, "he'll hook up his bar code wand to a computer. Everyday the computer will transmit back to headquarters a list of the packs, letters, and packages that arrived."

Even those specifics, however, still weren't enough to seal the deal. Marty knew that Gordon wanted more.

And so he ticked off the names of sellers of (1) bar code wands, (2) the little portable computers that stored information from those readers, (3) the scales, (4) the material handling system, and (5) the sorting machines. Marty even addressed details such as how big the portable computer's random access memory should be. And within a few days, he brought in cost estimates.

"Not bad," said Gordon. And so Marty's company beat out two rivals—slouches that had simply questioned Gordon about his needs and left without supplying a solution.

In fact, a month later, all the competitors did was hand Gordon some fuzzy concept papers telling vaguely how they would do the job. They didn't plunge into the specifics. Let's just say they practiced multidefect marketing.

Marty, on the other hand, was acting perfectly in line with the Friedman Formula and other concepts of this book.

For instance, he:

1. *Lived up to the slogan on the Avis buttons—tried harder.* Marty regularly visited his client and listened. And so he could sniff out the antelope that might have eluded the less alert.
2. *Micromanaged his people to get the job out on time.* Marty didn't let the little details slip by.
3. *Nullified Gordon's fear of failure by specifically telling how to track the all-important letters and packages.* Marty hadn't any choice. If the solution didn't work, Gordon might be fired.
4. *Showed a sharp competitive edge—the ability to move quickly and decisively, especially on the major issue of price.* Without *any* cost estimates from the rivals at this point, Marty's client

hadn't any idea what the numbers would be. And notice something else? At no time did Marty act like a salesman and roar how great his company's services were. He just buried himself in his potential client's problem.

5. *Behaved morally.* Marty honestly described what he could do. He didn't exaggerate the capabilities of the system he had in mind.

Follow the Formula, and Marty's example, and you'll more easily be "On Track with the Marketing Engine," discussed next.

Do's

1. Go after the business rather than waiting for it to come to you. Prepare for war.

2. Overcome company weaknesses that will give your competitors ammunition to shoot you down. Don't let them outsmart you on prospects you are pursuing. Of course, the investment in time must be worthwhile.

3. Track all known competitors and determine their market shares and influence in a client community. Collecting data on their strengths and weaknesses will better prepare you for each contest.

4. Use the results of competitive contests to gauge your capabilities, strengths and weaknesses. Pay attention to your win/loss rate, your profitability, and feedback from clients.

5. Anticipate and control those conditions that represent potential for failure—such as (1) unexpected calamities like computer breakdowns, (2) incomplete information on clients, (3) lack of internal marketing drive, (4) poor proposals, and (5) inadequate amount of organizational support.

Don'ts

1. Don't follow the "Publisher's Sweepstakes" syndrome. Marketing isn't like filling out the entry received in the mail and believing that the odds are in your favor of winning millions when you send it back.

2. Avoid the waiting game. Don't wait until the last minute, just before a solicitation is announced, to start pursuing new business. Your rivals will already have done their homework. And you'll spend too much to catch up.

3. Don't forget the unseen competitor, still lurking about. His proposal and influence, while invisible to you, are still influencing the buyer's decisions. So always act as if competitors are a step ahead of you.

4. Do not let creeping indifference—to client needs, marketplace strengths, or anything else important—do you in.

5. Don't shotgun the marketplace trying to find business where none exists. See first if there are any targets. And restrict the scope of your approach.

CHAPTER 3

ON TRACK WITH THE MARKETING ENGINE

This $17 million opportunity might have been overlooked by an inexperienced marketer.

It was a meeting of doctors, nurses, and osteopaths on a federal advisory committee discussing a proposed series of changes in the management of the Medicaid program. Thirty people watched from the audience. Frank Hanson, a crack marketer with a large computer services firm, saw bureaucrats, insurance company lobbyists, and others—but none of his rivals.

That was their loss. By understanding the doctors' concerns, Hanson himself would be better able to write a proposal for the computer system that the new regulations required. His actual role, in fact, went far beyond this. At the meetings, Hanson raised his hand and made suggestions when the committee asked for audience reaction. He also lunched with the members. "We like the interest you're taking in our problems," said one. "We're wondering if you could advise us on new rules. They don't teach management information systems in medical school."

Come bid time, guess who won the $17 million contract to operate a Medicaid information system?

The doctors themselves didn't select the bidders; they had just served in an advisory role, recommending rules and objectives for the Medicaid program and offering their two cents' worth on management issues. But the actual buyers, the bureaucrats with the money to spend, had learned soon enough who Hanson was.

Hanson had wired himself in. The doctors and bureaucrats saw him as a peer on whom to rely for advice. Hanson, a lion, was an honorary antelope breeder. This animal analogy has its limits, however—he didn't prey on the taxpayers. Rather, Hanson spent hours and hours listening to the doctors and bureaucrats so he could help them come up with the best solutions for managing the Medicaid program.

From the very start, Hanson was practicing a different kind of marketing, distinct from consumer product marketing and sales. His kind of marketing blurred the lines between his company's technical divisions (the ones creating the solutions) and those responsible for bringing in business.

Here, I'll tell how you, too, can develop a powerful marketing system and use it correctly. You'll find out about:

1. *How marketing differs from sales.* You're not a huckster trying to convert products into fast cash. Rather you're carefully catering to customers' needs.
2. *Nine reasons why the marketing of high tech services is special.* Clients' needs arise suddenly, and you usually can't use off-the-shelf solutions. It isn't as if you're marketing soap to people who bathe everyday.
3. *Seven marketing philosophies to avoid.* An example is "Fit and Start Marketing," when you fire the marketer as soon as he brings you the business.
4. *A marketing philosophy to adopt.* I'll tell you more about the philosophy that helped Hanson win the $17 million contract.

You'll also learn the usefulness of setting up a marketing organization. And you'll find out how to nurture a first-class marketing effort and successfully make contact with clients and promote your company.

Above all, you'll understand:

- The importance of integrating marketing into the jobs of technical people.
- Why marketing should continue beyond the lead through the life of a project.

HOW MARKETING DIFFERS FROM SALES

Wisely, Harold J. Leavitt, a management and marketing expert, has said "Selling is preoccupied by the seller's need to convert into cash. Marketing is preoccupied by the idea of satisfying the needs of the customer by means of the product or service."

An even better-known guru, Peter F. Drucker, goes further.

In his opinion, marketing ideally will make selling super-fluous; just know your customer, make the product or service readily available, and it will sell itself. He even says marketing and sales are antithetical.

Consider this case history:

> Once, not even P. T. Barnum could have won a multimillion-dollar computer security contract with a hospital in central Pennsylvania. Then news stories appeared, telling of young hackers prying into computers.
>
> Now the vice president for the hospital's computer department had a problem and was eager to find a solution.
>
> "My job's on the line," Ed Ferris said after calling me in. "If one of those kids gets into our system, *we'll* be the ones in real trouble."
>
> "How's that?" I asked.
>
> "The government makes us protect the privacy of patient records," he said, "and we're wide open because the hackers can dial us up. We're just using ordinary phone lines." What's more, by chance, a programmer had caught a colleague on the verge of a possible embezzlement by way of the hospital's computer system.
>
> And so Ferris and I did some business.
>
> Beforehand, I hadn't sold. I'd marketed. My marketing program was in place long before the vice president realized he had a need. I hadn't any idea what the exact problem might be—only that someday there might be one. And thanks to personal contacts, Ferris knew who to call for a solution.
>
> "Ah," some might say, "so you were a salesman—ready with prospects, already softened."
>
> No, a *marketer*.
>
> I didn't have a supply of 100 security systems to unload to scrape up cash to build more; instead I was trying to fill the hospital's acknowledged need. And it was one-of-a-kind need. Years might pass before Ferris wanted another security system—

if ever—unless the hackers grew smarter and he wanted add-ons or a state of the art replacement. I'll be ready for *that* special need too.

Clearly, in Pennsylvania, I was *marketing* a *solution*.

As I've said, a single event or series of events had whetted Ferris's appetite for my company's services. The solution, moreover, didn't sit on the shelf at the local A&P. And Ferris couldn't really compromise on either the solution or the price. Only a limited range of possible alternative solutions would have been available even if he had opened the project up for bids.

Granted, John Glenn once joked: "How would you feel if you went into space on a craft built by the lowest bidder?" But normally services bids don't vary that much in quality or even price. And Ferris realized this.

So it wasn't as if he were eyeing the different price tags and packaging of 50 brands of soap. He knew my true role. I was solving his problem (computer security) rather than addressing an ongoing human need (cleanliness) that would mean repeat business in the usual sense.

Ironically, by acting as a salesman rather than a marketer, you might actually endanger future contracts with old customers. So don't push your "favorite solution" on someone who doesn't need it.

That may be true even if you're selling a product instead of a service.

Thanks to the wiles of a salesman working under quota, some people at the Department of Defense may think long and hard before buying another large computer from International Monolith. The salesman sold the Monolith Model 970 as being able to serve many users at once—"multi-tasking," in computer industry jargon. "Serial-tasking," though, would have been more on target. The machine just wasn't powerful enough for the job, and programmers had to toil two shifts a day, six or seven days a week, to meet their deadlines.

Not all sales reps are so callous toward their customers, but in this case the man was clearly more sensitive to his employer's cash flow requirements than to Defense's needs. Don't think that his victims will be amnesiac.

At least the sales rep was acting competitively. So many people in high tech services don't.

NINE REASONS WHY THE MARKETING OF HIGH TECH SERVICES IS SPECIAL

Summed up, here is how *marketing* high tech services differs from *selling* consumer products, which normally meet *steady* needs:

1. *Client's needs—or at least the perception of them—*suddenly *reach the level where the buyer is interested in the solution. Your buyers don't keep a stock of solutions available to be used when needed; they can't load up on solutions as they would soap or toilet paper.* New needs can come from changes in areas such as:

- Government regulations.
- Technology.
- Economic conditions.
- Internal factors like security, costs, and productivity.

Your buyer will convert those needs into reasons to buy, and you must be ready. Don't tarry. As soon as you decide that your clients may have new needs, start thinking about possible solutions. You never know when the phone will ring with an Ed Ferris at the other end.

2. *The best solution for those needs doesn't always exist off the shelf—at least not before you implement it.* One source often won't do. You may have to plow through two dozen equipment catalogues or study 50 articles in trade journals. For the hospital's security system, I bought equipment from five suppliers. That's typical. What's more, even when services companies locate the right combinations of hardware and software, they often must "tweak" them to meet the client's exact needs.

3. *Your best "solution" may differ radically from the one your rival would have proposed. You aren't just marketing a different brand of shredded wheat.* It's as if you're designing an office building. Yours may look different from your rivals'. Your services client, however, won't care—as long as your answer is inexpensive, meets his specs, and works to meet *his* needs.

4. *The means of production for solutions may be available but not always operational.* Until we installed our tailor-made security system at the hospital, the "best solution" didn't exist. I knew my company could put together the system. But at first it was only a vision inside my head. I had to assemble the right combination of hardware, software, and people to help blend everything into a good solution.

5. *If you must modify your solution, you needn't worry about changes in production and distribution.* You're only choosing the *components* of the solution. You're not General Motors retooling its production line to make larger or smaller cars. If I'd found that the computer gear for the security system didn't meet the client's performance specs, I'd have fiddled with my software to speed it up or ordered a new computer. I wouldn't have had to tear apart a production line.

6. *Normally you needn't have distribution channels throughout the United States. And it's often easier to go international.* Price, speed and quality will almost always win out over location.

For instance, J. Michael Nye, the president of a computer security firm in Maryland, works out of a small recession-racked city where wages and living costs are lower than in, say, Washington, D.C. The name of his firm is Marketing Consultants International. And Nye likes to joke, "In Japan, people in our specialty don't think of the phone company when they hear the initials MCI. They think of us."

Thanks to Federal Express and lower long-distance rates, a high tech company in the Silicon Valley can program computers for an insurance firm in Massachusetts. The California people might even tinker with the software by way of the phone lines.

Please note that there are exceptions. For political—not technological reasons—a state or city government might favor a services company with the right address.

7. *Logistics and distribution usually aren't as critical as in the consumer market.* Teamsters can strike in New York. Hurricanes can flood Florida. But your work usually can go on, since your people and services operate mainly at local client sites.

8. *Frozen orange juice may sell for $1.30 for a 12-ounce container, no matter who's buying it—but the prices of services are subject to negotiation and may vary from customer to customer.* Prices will also be sensitive to competition. Interestingly, because the labor and equipment costs of the services companies are so similar, their initial bids might be almost identical in some cases.

9. *You're interested only in people and companies with specific needs and interest in meeting them.* You're not commissioning demographic studies of millions of potential orange juice drinkers. You're marketing only to a select group—customers who (1) don't already have solutions in operation, (2) know they have problems needing solutions, (3) can afford to change their ways of doing business, and (4) can afford the solutions.

Ideally, of course, in both selecting markets and servicing them, you'll act just as competitively as do the suppliers of consumer products.

Also, you should avoid seven destructive marketing philosophies that so many high tech companies mistakenly adopt.

SEVEN MARKETING PHILOSOPHIES TO AVOID

The seven are:

- "Fit-and-Start Marketing": Fire the marketer as soon as he brings you the business. Don't invest in the future. [*Remember: I'm talking here about philosophies to AVOID.*]
- "Vanity Marketing": Toot your horn and ignore clients' needs.
- "Marketing by Wandering Around": Act as if you're a huckster paying regular visits to sell a hot consumer product.
- "Cul-de-Sac Marketing": Depend too much on a niche.
- "Nomadic Marketing": Don't confine your efforts to your strong areas—or even to those in which you're mediocre.
- "Limping-Giant Marketing": Use size to overwhelm rivals, without appreciating the peculiarities of the services market.
- "No Push/No Pull/No Click Marketing": Don't include clients' needs in the planning process and don't fight hard for new and expanded business.

Not every company will use fit-and-start marketing at all times. Sometimes it might instead sin in another way, such as through vanity marketing. And on occasion the same firm might even be marketing properly.

Styles constantly shift. Such changes might themselves suggest a problem: incompetence and a lack of direction.[1]

Do you remember the old saying that God takes care of drunks, little children, and the United States? Well, perhaps, He also protects some inept companies in professional services. It's amazing how they survive. Why? Another explanation is more temporal: the Old Boy network.

But while God endureth, the Old Boys retire and die, and even He can be fickle at times; so the true professionals in marketing should depend on neither divine nor bureaucratic intervention.

Otherwise they'll very likely see rivals outsmart them and steal their antelopes. Morale will decline. Then outsiders may buy them out. Just look what happened to Philistine Tech, where the engineering division alone was losing several million dollars a year. Yet another offender—a fit and starter—was Katydid Systems.

Fit-and-Start Marketing

Don't desert your marketer, fit-and-start fashion, if he succeeds. Don't be another Katydid.

> Katydid had been in data processing services two decades, with a bread-and-butter income from a steady group of clients. Then one day, the biggest client, an import-plagued car maker, said it would drastically cut back the use of outside contractors to handle all routine computer matters. Within weeks another client, in aluminum, made a similar announcement. Ken Katydid gulped down enough tranquilizers that week to stock a small pharmacy; the two lost clients had accounted for more than half of his company's income.
>
> He immediately hired an aggressive marketing man, the first of his breed at Katydid Systems. Earlier the top executives themselves had marketed as a sideline.

---

[1] Of course, styles might change simply because the market itself does.

Within two years, the marketing man more than made up for the lost business, and he was about to launch a major campaign to raise sales 50 percent within two years when Ken Katydid ushered him into the presidential office.

"Nice job," said Katydid, "but we're going to have to let you go."

"But I'm getting new business."

Katydid nodded. "And also giving lots of business—to airlines, hotels, and other places where you spend your money. See, I've got to cut my overhead costs. Those next five big prospects—I want to keep my prices competitive. We're swimming in new business. No need to keep paying you and the airlines."

The marketing man's parting words were, "Just wait a few years and you'll be aching to make up for the income you lose when your present contracts are up. Only, this time I won't be around."

He was right. Katydid Systems scaled down its marketing efforts—its investments in the future—while its rivals grew stronger. Within two years the grasshopperlike company was on the verge of bankruptcy.

A slight variant on this scenario happens when a firm hires a marketer for one or two years to expand business, rather than simply to make up for lost contracts. Here, too, the company fires him as soon as he meets goals. Contracts expire. Then the company may discharge 50 or 75 percent of its staff.

Yet another fit-and-start variant is when a company discharges a marketer for the opposite reason—because business has fallen off and the top executives think they can do his job themselves. Maybe the decline *was* the marketer's fault. More likely, however, the technically oriented executives didn't understand business management practices and philosophy; perhaps they saw marketing as an ancillary function or as a remote staff function, a vestigial remnant of their days working for a product manufacturer.

"I can do the job with my pinkie," the top executive may growl. "It's just one milk run after another. I'll just fit it in between my normal things."

The fit is bad. If the top executive appreciates the demands of marketing, he finds he must reduce the time for other work; so the remaining contracts suffer lack of guidance from him. Or the other problem occurs. He can't squeeze into his schedule enough time for marketing.

Even a part-time marketing consultant on retainer might save such a company the embarrassment of poor financial statements and layoffs of their star professionals. Ideally the company will wise up. Marketing is an integral part of its business—not just a way to grow but a way to survive.

Vanity Marketing

A prime example of vanity marketing—appealing to someone's corporate ego rather than to your client's needs—is Crowe Research.

> Crowe Research might as well have been selling Rolls Royces. Fred Crowe, the holder of an engineering degree from M.I.T., overcompensated for his lack of a true business background and came across as a huckster going after the snob market.
>
> "Our solution," he loved to tell clients, "is the only one. We're the best. Our competition doesn't even come close."
>
> Potential clients' specs were irrelevant. Fred *knew* what was best for them.
>
> Of course he didn't offer reasons for the buyers to select him. Contracts officers and project engineers might have spent months preparing solicitations, carefully outlining their needs and preferred solutions; but Fred disregarded all—in favor of his Rolls-level solutions. Unfortunately, the clients wanted Chevys. Or perhaps Mercedes. Fred was a brilliant engineer, but he forgot that the most expensive technical solutions weren't necessarily the most cost-effective ones. Maybe the buyer couldn't afford the very best. Or perhaps even the very best wouldn't fit in with the client's mode of operation. Fred shut his eyes and ears if a prospective client insisted on a mediocre solution. Only the optimum one counted. And so Fred received flattering write-ups in technical journals—but a fraction of the business he'd expected.
>
> At times, of course, Fred did win contracts. Sometimes he successfully convinced clients that he could solve problems. Commonly, however, it was because the buyers simply hadn't done their homework; so they accepted Fred's advice. Or else Fred used some chicanery, political or otherwise.
>
> In the mid-1980s, new technology appeared. Fred's solutions seemed less attractive than ever, and only strong pressure from clients forced him to change his ways in time. Such scenarios happen often. You can't turn around a company like Crowe from

within—at least not without a Fred himself on your side. It takes pressure from the buyers, loss of major contracts or the purchase of rivals (if the money can somehow be scraped up) for the vanity marketing firm to see the light.

Vanity marketing *hurts*—both the balance sheet and goodwill among clients. It can be the express route to bankruptcy court.

Marketing by Wandering Around (MWA)

Your marketers can't win work just by visiting X number of prospects. Walk in unprepared and just a few will buy. Don't imitate Sam Swiftt.

Swiftt, a marketer with a big computer services company, lived up to his name, winning marathon after marathon in college. He loved to run. He loved to walk. Imagine the glee Swiftt felt on picking up some old sales management textbooks that rambled on and on about the need for reps to see X number of possible clients each week. Sam was ecstatic when his marketers handed in long lists of client visits.

Eagerly his people toted up the possible income from such prospects; then they turned the names over to a technical department for follow-up and closing deals. Alas, the approach was folly. The technical departments were already swamped and couldn't spend time on the care and feeding of these prospects. They faded away. And so Sam leaned on his marketers to visit even more prospects. He treated them as if they were sales reps on the road moving Laser Tag games at the height of the craze—as if clients were crying out for more goods because of Extra Demand.

He actually ordered his people to ask: "Do you have any work for us today?"

The pickings were tiny. Less than 1 time out of 100 was the reply "Yes," and usually the contracts were minnows.

Sam had forgotten that the work won't come automatically when you ask for it. It's there when the client *needs* it: when he wants a new technology, or enjoys more funding, or reads about problems such as computer fraud. Swiftt became a laughingstock. His rivals called him Sam Slow.

I felt sorry enough for Sam to put him on to what was happening—he wasn't a direct competitor.

"People either dole out business to their favorite firms or they'll go through an extensive competitive solicitation cycle," I said. "Don't waste your people's time on one-minute social calls. And give 'em a chance to get to know their new prospects gradually, so you'll be on the list when there's a problem to solve. Above all, give them time to learn their clients' needs, and to research solutions. And to tell why your outfit's the best choice. They can't do all this with a list of 30 clients, so don't make them rush the job. Wooing a client follows the rituals of dating. You don't say, 'Glad to meet you. Would you sleep with me?' Of course you might be able to find prospects with immediate needs. But if it's the big contracts you're after, you'd better think beyond the quickies. Suppose your client says he might need your solution a year or two down the pike. But your people won't have time to cultivate him. And when he's finally ready to do business, you'll find your competitors swarming around."

"Are you calling me a basket case?" Sam angrily said. "What about that $6 million contract with Hank Johnson's outfit?"

"And what about the fact that you and Hank went to Penn State together?" I said. "You didn't get that contract because your people saw Hank every week. You got it because you two were in the same old boy network. Frankly, I think Hank hurt you. He made you think your present methods were working."

Alas, blind to everything but the flukish growth of the contract with Hank Johnson, Sam continued his MWA approach. Then Hank moved on to another job. His successor, with an old boy net of his own, cancelled Sam's contract, and soon afterward, two other major clients failed to renew.

Clients whispered among themselves. "This guy's a joke," said one to a friend. "How can he serve us if he doesn't even know how to market himself?" One of Swiftt's nastier rivals suggested that as long as Sam was so keen on running the marketing department based on sales-call quotas, he might as well give his marketers vacuum cleaners to sell at the same time.

Cul-de-Sac Marketing

Cul-de-Sac Marketing—selling to a little niche—can be just as hazardous to your company's health as is Marketing by Wandering Around. Unforeseen changes can do you in. You should diversify—break into related fields of business—to survive. Beware of Ted Jenkins's fate.

Jenkins, who had toiled long and hard for nonprofit think tanks ranging from the Rand Corporation to the Institute for Defense Analysis, decided to form his own services company. He was a leading authority on the way nuclear war would wreak havoc on radio communications. And now he expected federal contracts to flow his way. They did. War III communications, however, was the company's only specialty.

Ted and his partners never branched into other communications-related areas. Nor did they enter other aspects of postblast military planning. And he wouldn't let his professional staff (assigned to a contract) do any marketing or proposal writing—his obstinacy being a surefire way to crimp expansion plans.

For a while Jenkins's company prospered, relying mainly on add-on work to current contracts, while inflation bloated earnings figures. But in effect, his firm had plateaued. He might as well have been circling in a cul-de-sac.

Then a new administration came to Washington: one sensitive to campaign contributions. Ted may have been the best authority in his field. But the second best guru, meanwhile, had left a famous think tank to form a company of his own; and along the way he'd built up some valuable political and bureaucratic contacts, who, as soon as possible, shifted business away from Ted.

For a nuclear war communications expert, Ted showed scant appreciation of the age-old principles of redundancy and contingency planning. All his contacts were in the area where his rival had usurped him. In this case, Brand X bidder's influence did Jenkins in; but the problem instead might have been someone with lower prices, or new technology, or perhaps another complication such as the resignation of key staffers.

Whatever happened, Ted hadn't allowed for the vagaries of his market. He had not prepared for changes in technology, his competition, and the economic and political climates. And so he lost all. Soon another company would move into the empty offices, the site of so many late-afternoon bull sessions over Doppler shifts and electromagnetic fields and digital communications in the year 2001; but right now they were desolate—Ground Zero or the financial equivalent of a neutron bomb.

Nomadic (or Meander-Style) Marketing

Some high tech services firms act as if Bedouins are in charge, as if they must constantly pitch their tents in a new part of the desert. In fact, they soon will be in a desert: one of their own making. They haven't confined themselves to specialties where they have the best contacts and expertise, and prospective clients soon catch on. Like the Cul-de-Sackers, the Nomads lose out in the end, but for the opposite reasons.

> Don Lawrence was a nomad. His company pursued business in areas ranging from solar cells for Star Wars to information systems for the Social Security Administration. He jumped at every solicitation.
>
> "Why not?" Lawrence would counter skeptics. "Everything's a system, isn't it?" That was his favorite word. "A good systems man can do anything. You can always hire experts. But it takes a systems man to get something out of them."
>
> Alas, the wells in Lawrence's desert soon dried up. With so many fields to worry about, and with skimpy credentials in most, his company had trouble finding and keeping the best specialists. Only the mediocre would work for him. What's more, by going after jobs for which his firm was clearly unqualified, Lawrence raised questions about his own competence.
>
> Lawrence still had enough capital to try to buy out some smaller companies with good reputations in their niches. But those new people, especially the marketers, soon left. Why shouldn't they? How can a nomadic leader keep his larder full when he's violating the most basic rule of competition: Apply your resources to the work for which you're best qualified. In the end the sands will bury him.

Limping-Giant Marketing

You may be a leader in appliances or autos—or even computers—but that doesn't assure you success in the services business. You can't live off your name in other fields. Omni Electric discovered this all too painfully.

> Omni made TVs, radios, electric coffee pots, refrigerators, washing machines, dishwashers, generators, industrial motors, just about any product through which electrons flowed.
>
> Bill Alexander, the head of a major Omni division, proposed that the company enter the services business.
>
> "After all," he said, "we have all those scientists and engineers on our hands. They're quite capable of doing engineering design

work. Including the kind we normally subcontract out to smaller engineering services firm. So why can't we sell their services to our existing client base?" An executive with a large construction management company—or other firms employing people useful in the services industry—might have had the same thought.

Omni's president loved the idea. Alas, however, Omni was a poor lion and impossible Snark player; it was new to the high tech *services* game and didn't know how to outsmart rivals through precontract marketing and good proposal writing. Nor was its overhead as low as the competition's.

Alexander's response was to step up direct mailings to buyers of Omni generators and other industrial or business products.

"Here is your chance to go Omni," a typical letter said, invoking The Name. "We have moved into services, too, everything from computer programming to security systems, and we will offer you the same Omni quality you have come to expect in our products. As a Fortune 25 corporation, we will bring new stability to the engineering and services industry."

Prospective clients, even the existing customers of Omni's *products,* still balked. Omni was not among the leaders in the areas where it proposed to offer services. And its professional marketers were busy pushing not just services but also generators and other hardware. Omni's services marketing didn't fail; but in this area the company merely limped along, humiliating Alexander, who had promised his bosses quick results.

Perhaps Omni would one day be successful, but not until it had dug far into its deep pockets. It was an expensive education as Omni learned to cut costs to compete with other companies and tailor solutions to clients' individual problems. The company might have saved itself much grief by setting up a new marketing operation for professional and engineering services, rather than making do with the existing people and methods.

No Push/No Pull/No Click Marketing

Don't get so wrapped up in strategic market plans that you ignore the needs of individual prospects at the moment. Respond to the *push* of those needs. And be prepared to *pull* quickly for new business. Then your marketing efforts will *click.*

UniVu Systems prided itself on planning and, in fact, outwardly enjoyed at least limited success.

The company had penetrated a cross section of the commercial marketplace with a fairly steady, if moderate, growth. It even offered computer services to several offices within the Pentagon and started 20 regional offices, from Maine to Alaska. UniVu's income grew steadily if not spectacularly. And it enjoyed respect from clients, especially the federal ones; bureaucrats admired UniVu's inexhaustible supply of statistics on the fate of America's technology and the world's. The company's mainframe whirred away day and night, identifying new areas the company might penetrate. UniVu was brillant at spotting growth markets, whether it was solar cells during the 1970s or Star Wars lasers during the 80s.

UniVu, however, failed to carry out its plans fast enough in those and other growth markets. It couldn't quickly identify the needs of *individual* prospects.

People at UniVu hated to mess with such grubby details as when the Department of Agriculture might put out a request for proposal (RFP) for an information system design, or when the Bureau of Labor Statistics might want a giant local area network (LAN). They were too busy analyzing broad trends and thinking in the long run. UniVu forgot one of the better observations of John Maynard Keynes: "In the long run we are all dead."

And so even if UniVu noticed individual opportunities, it all too often failed to zero in on the prospective clients' exact needs and consult with them regularly. The company didn't fight for new or expanded business. UniVu's marketing efforts, alas, were often too weak and too late.

To be sure, UniVu *was* growing. But it hadn't grown nearly as much as such a respected company should have. It won an average of one nine-figure contract for every 10 tries, and often the victories came when there were few competitors, perhaps because the rivals thought that UniVu was wired in. Even UniVu's friends among its clients felt frustrated at times. When just one bid arrived, a buyer felt like a failure; he had no other place to turn for a better deal. Of course, in competitive situations, UniVu almost always lost—sterling reputation or not.

Too bad UniVu's vaunted mainframe didn't analyze what was really happening: This no-push company wasn't pulling, either. It didn't follow important zero defect marketing procedures such as locating a prospective lead; befriending the buyer and his organization; gathering the necessary marketing intelligence, including the client's less obvious needs; influencing the statement of work and the solicitation, if possible; maybe providing some free position papers; searching for competitors; and looking for prospective subcontractors that had experience with the part of the government (or corporation) involved. UniVu might sink millions into computer equipment for itself. But it wouldn't favor *individual* marketing efforts with the necessary time and money.

Granted, UniVu tried in its own cumbersome way. It mailed glossy brochures to scores of prospective clients; even the trusty mainframe did its part, sending out megabyte after megabyte of ballyhoo via MCI Mail to computers at other companies. And expenses were indeed huge. Wins cost 50 cents of every income dollar in the search for new clients.

In a nutshell, UniVu's problems were:

- The company let its plans drive the business, expecting that the designated markets would search it out.
- It lacked an action program to select and pursue immediate prospects suggested by strategic plans.
- UniVu didn't respond to changes in the marketplace and in client needs. It didn't shift around people and resources accordingly.

UniVu did not, and could not, make anything happen. It couldn't pursue clients. It couldn't close deals. It could only plan.

A MARKETING PHILOSOPHY TO ADOPT (AND THREE DIFFERENT VARIATIONS)

How to avoid the mistakes mentioned above?

Remember an eternal truth: *In high tech services, marketing functions first of all as the front end of any contract.*

This underappreciated activity doesn't just bring in the business. It's also responsible from the start for determining and laying out all the technical, management, staffing, and cost solutions for clients' approval. And this happens *before* the awarding of a contract.

Marketing isn't any different from what would happen if you needed to solve new problems in the middle of a contract. In that case the project manager would approach the client—that is, market him—with new solutions; he would be seeking expansion of the contract. Bear in mind that marketing is a continuous thread. It goes from a lead to a proposal to closing the deal to the execution of the contract. You market to get business, keep it, and expand it. You use it to survive and grow.

Above all, you must integrate marketing with the technical side of the business. You can't segregate it as sales is segregated from the engineering side of the products business. Your marketing campaign will fail if you operate it like Swiftt's marketing by wandering around. Marketing should be the front end of your projects. It should determine how you carry out contracts, what solutions you adopt. You've failed at blending marketing into the rest of your organization if your techies exclaim, "But I am not a salesman." You've succeeded if techies and managers regularly team up with marketers to visit existing and prospective clients, in hopes of winning new business.

Three Different Ways of Setting Up a Marketing Function

If you decide to set up a marketing function, you have three choices:

1. A centralized marketing department.
2. An integrated marketing function within a technical department.
3. Letting technical managers market.

Centralized Marketing

The centralized marketing approach works this way. The marketing department supports the technical ones and the company as a whole. The techies pursue more work from existing clients, while the marketers go after new ones.

The marketers also:

- Get the company into new kinds of ventures.
- Promote the company and the company image at professional and industry association meetings.
- Develop technical brochures.
- Work with the technical staff to develop marketing plans of action to go after specific bids.

Normally a central marketing department will report to the chief executive officer or the chief operating officer. The department is key to the whole operation.

Within the department, staffers receive assignments of client cases or accounts in several ways:

- Client organization. The marketer may market just one client organization and represent all the technical capabilities of his companies.
- Technical specialty. The marketer might deal with more than one client but confine himself to the technical areas in which he shines.
- New venture area. The department assigns him to pursue new venture areas and market segments.
- A combination of the above.

In deciding which marketer handles what, you might consider not only experience and other qualifications, but also past contact with clients.

Integrated Marketing Function
In integrated marketing, you assign marketers to each technical group in the company. The benefit here is that the technical group enjoys the services of a full-time marketer who woos business across many client boundaries, within the group's specialty. He reports to the group's manager.

Disadvantages? A few. Marketers dedicated to different groups might find themselves fighting over clients and hiring staffers for similar jobs.

Letting Technical Managers Market
Whether you're a small, medium, or large firm, you can thrive with this approach under certain circumstances; but you're at the mercy of changes in the market and in technology. You must constantly make new client contacts.

Your main focus here is to let technical managers expand existing contracts and acquire new contracts from the same client organization.

To break into a new client base, you must negotiate with other technical divisions for new resources on a quid pro quo

basis. There are problems. Divisions must decide who will identify new kinds of client as prospects; who will reign supreme on
the new turf; and how to pay rival divisions for use of people and
other resources. Also, technical managers must enjoy relief from
direct management of projects, so they'll have time for marketing.

Still, many firms thrive this way. Some boast sales of more
than $250 million yearly.

The Advantages of Setting Up a Marketing Organization

A marketing *function* isn't the same as a marketing *department*.
Should you have a department?

Even if you plan to let your technical people help market,
you still should consider establishing a marketing organization,
however small. It might consist of just one part-timer. Helen
Edwards found to her chagrin that even fledgling companies can
profit from such services.

> Edwards, a brilliant alumna of a large accounting firm and a
> specialist in computerized audits, went off on her own. By now
> she should have been netting $100,000 a year at least. Instead of
> raking in the money, however, she's back in bondage to her old
> employer. Here's one reason why. Because of the smallness of her
> company Helen couldn't win big, long-term contracts very easily.
> She should have scurried constantly for new business to keep up
> her backlog once current contracts expired; but she was too busy
> working on existing jobs. If Edwards had hired a part-time mar-
> keter, she just might have survived.[2]

Her fiasco is just one of many. Over four fifths of small firms
in high tech services will die within two years. And in some
cases, a good part-timer just might have made the difference.

There are other arguments for setting up a marketing orga-

---

[2] Edwards could have hired the part-timer through an executive search firm or
through the marketers' old boy network. She could have paid him the equivalent of a
technical manager's salary (in proportion to the amount of time spent on the job). Plus,
she could have given him a bonus based on a percentage of the value of the contracts he
brought in. In some cases, in fact, a retainer might be an alternative to salary.

nization. For instance, what if your win rate has dropped below 25 percent or the board of directors wants you to grow? Or suppose you want to break into a new market niche . . .

How to Create a First-Class Marketing Department

Some of the major issues here are:

- The categories of people to hire.
- The importance of ranking marketing people alongside other senior managers or professionals.
- Making fullest use of the talent of the people you hire.

Whom Do You Hire?

You should recruit at least one of the types of people in the following three categories to help with the direct kind of marketing:[3]

- Category One: A Marketer.
- Catetgory Two: The Contact Person.
- Category Three: The Executive Consultant.

Category One: A Marketer
A marketer should be far more than a lead-gatherer. You aren't hiring him to bring in X number of leads in Y time; instead you want a professional who is interested in quality rather than quantity, and who can follow up important leads and perhaps even become a project manager.

You'll assign a marketer to an account, a government agency, an application area, or another market niche—or perhaps let him roam around as a generalist. In fact, try to hire marketing people who can pull double or triple duty. A talented architectural engineer, for instance, might specialize in designing buildings for the Defense Department that are hard for KGB

---

[3] The phrase *direct marketing* refers to focused marketing efforts—not simply marketing that technical people and others do as part of their regular jobs.

agents or mere burglars to break into. And he also might boast skills in ostensibly unrelated areas such as computers. Ask when you recruit. Maybe the architect has a knack for computer-related matters, having purchased an elaborate system for his old firm. You'd be surprised at the number of people who shine in dozens of areas beyond those for which they have formal credentials.

Don't neglect traits beyond versatility. Show a prospective marketer the door if he lacks either initiative or imagination. Value management and communications skills, too. Also make sure your marketer can understand the technology, plans, policies, procedures, and budgets of his prospects—and, above all, potential clients' needs.

Clearly, market engineers are prime candidates for your marketers. These marketers will boast professional degrees and the right track record, including successful project management experience.

Their special discipline includes technical, business, financial, and administrative responsibilities.

So consider using a market engineer to manage the entire marketing life-cycle—from lead acquisition, client liaison, solution generation, and proposal development, to closing the deal, executing the contract and expanding the current work.[4]

The marketer is the most basic category: someone who endears himself and your company to known prospects, winning their trust along the way. A successful marketer might nurture no more than five major clients at once—their care and feeding takes time. Don't distract him from this central mission. Do not overload his dance card with a list of new prospects to contact; otherwise rival companies may steal the customers he's carefully cultivated. You can help your marketer keep clients by hiring another kind of employee, the contact person.

---

[4] You could find the full-time marketing engineers through some of the same channels used to locate a part-time marketer. Here, too, you would pay the marketer the same as his technical equivalent. In fact, in this case, the marketer himself would be technically qualified even if he wasn't as specialized as the other technical people.

Category Two: The Contact Person
(Door Opener)

A contact person usually holds a senior position in the client industry. He enjoys respect. Perhaps he's a former marketer or senior technical manager or engineer or he has contributed to the technology. You might say he's the equivalent of a newspaper tipster, the one bringing in the hot leads. He keeps abreast of general conditions in the industry—of changes in plans, programs, people, and the economic climate.

Suppose headlines appear about a breakthrough in CAT scanners. The contact person might draw up a list of hospitals interested in possible help in installing or using such marvels. What's more, your contact person should be able to introduce the marketer to prospective clients.

Pay the contact person a retainer, or a commission based on leads brought in and contracts won, or perhaps a combination of the two.

Category Three: The Executive Consultant

Americans love Leading Authorities. Newspapers quote them. Professional associations honor them. You're much better off with a Leading Authority on your side: a gray hair to whom people will listen for advice—ideally in areas where you'll be hustling for business. With a Leading Authority helping, in fact, your marketing efforts will seem more genteel. He'll be your company spokesman at meetings of presidents of corporations and yacht clubs and say, "These guys (fill in your firm's name) are tops in their field with a new technology." And the president or yachtsmen will nod and think, "Jeez, let's think of (your name) next time we have a problem related to (your field)."

Above all, you'll be more successful because Leading Authorities know heads of companies. If you can somehow hire Lee Iacocca in his retirement, he might introduce you to good friends at Chrysler or Ford whom you can then market. And he'll be more than a door opener. He'll say: "This is my good friend (fill in your name) and he's trustworthy and worth your time."

Of course, your Leading Authority won't be a mere marketing or contact person. The jargon is "Executive Consultant."

WHAT A MARKETER SHOULD DO

In short, a marketer should manage or participate in:

- Identifying the client and marketplace demands, needs, and opportunities. The marketer should receive help from the whole organization, and, in fact, the best companies will insist that technical people help him do other tasks on this list.
- Determining the likelihood of leads and other opportunities actually panning out.
- Discovering—and assessing the threats of—competing products and services.
- Doing the same for threats, such as harmful legislation.
- Considering the firm's strength and ways to overcome weaknesses.
- Qualifying leads to determine client needs and buying reasons.
- Planning and developing good solutions for the client.
- Also planning better management strategies to beat the competition, while working within the client's cost expectations.
- Selecting and negotiating with suppliers and subcontractors.
- Directing organizational responses to meet various challenges—such as bringing in people from your company or the outside: technical people, research and development (R&D), legal, a number of categories.
- Managing proposal development and developing strategies for winning.
- Determining project costs and the winning price.
- Making sure that Murphy's Law[5] stays out of the way.
- Closing the deal; contributing to contract negotiations.
- Following up contract awards to determine if additional work is available.
- Searching for new ventures, market segments and niches to pursue, and drafting plans of action.

Two of the most important functions of the marketer are, of

---

[5] Explained on page 129.

course, client contact and promotion—discussed in the next two subsections.

Using Client Contact to Move Ahead on Track

If you're a marketer, you won't just talk to a company or government agency while you're doing the duties listed above. You'll be communicating with *people*. How do you gracefully find out a client's needs and persuade him to sign a contract? Here's a summary of the steps to reach this goal.

1. *Get your prospect talking to you freely—about anything.* Let him talk about his hobbies, lunch, a mutual friend, the weather, a solution to the national deficit, anything, especially mutual problems. But let him talk. You're trying to build rapport.

2. *Next find a way of redirecting the conversation so you can begin talking about the client's needs.* Discuss his job, his problems, and constraints. Show you understand the demands of his work, then draw out his needs related to the complications.

3. *Clarify the situation, summarizing the needs of the client and what beneficial solutions you might offer if he is willing to accept your answers.* Immediately offer your tentative solutions and how you'd effect them. And if you don't know of a solution? Then make an appointment to bring your best techie to discuss it as soon as possible. Or you can offer to transmit it the next day.

4. *See if you can get your client to make a commitment.* It needn't be on the order of, "Hey, let's sign a contract," but rather: "What do we do next?" Either suggest the next step and win his agreement or ask him what he proposes for the next step.

You both must make a commitment as to what the next action should be, and when it should happen. For instance, the commitment might be that you'd propose a draft statement of work or a position paper on the necessary system architecture or whatever else the client requires.

Through written material and otherwise, you can keep the relationship alive—since it might be yet another six months or a year before the client can issue a solicitation or sign a contract.

Perhaps you can discuss staffing requirements and additional details.

5. *Get your client to introduce you to others in his organization.* When he does so, he commits himself further—by offering you and your solution as worthy of discussion. This will also give you a chance to discuss your solutions in greater detail and perhaps modify them if the buyer or anyone else shows skepticism.

6. *Try to close the deal.* Informally you may have come up with a package and said, "OK, now here's the outline of this package of how things are going to be—there's nothing more. If this flies, we'd like to be your contractor." Be positive!

If your prospect says, "Maybe, but I'd like some competitive bidding," you should not give up. If nothing else, your sustained contact with him shows both confidence in your abilities and a commitment to the proposed job. At this point you haven't even discussed price yet. You remind him, though, of the investment in time and money that you've put into the possible project.

Whether your solution is a technical or a managerial one, you reiterate how successfully your company can do the work. Don't tell. *Show!* Sum up again why your answer should be irresistible. Make this job competitor-proof by showing your company's uniqueness. Reduce your prospect's fear of failure. If need be, you should expand on the benefits of your solution and, while you're at it, see if you can uncover other areas of need.

Is your client still showing some hesitancy? Then see if either you've misunderstood his problem or he's misunderstood your solution. Rephrase your description of the solution. Ask questions to gain yet more insight into his problem and why he's still somewhat skeptical.

How Your Marketer (or You) Can Promote Your Company

You or the marketer shouldn't just think in terms of the present contract—also encourage present prospects to give you more business. And whet the interest of future clients. Promotion is one way to keep the marketing engine moving ahead to new deals. Here are some tips:

1. Don't fall into the trade show trap. Your services are usually too complex to sell, and the atmosphere too circusy. Go from time to time, but avoid tipping off competitors. Keep your mouth shut and your eyes and ears open.

2. Go to professional conferences and use other occasions to befriend your prospects and make them comfortable with you.

3. Present well-researched professional papers there.

4. Publish professional papers.

5. Have glossy promotional literature and capability documents to hand out at conferences and also to mail to selected prospects. Organize material logically. If you're doing a capability report, list your accomplishments by functional areas, such as security, communications, architectural engineering, and human services.

6. Generate publicity in trade publications. Your goal isn't to get on the evening news. But why not seek selective publicity in trade publications and maybe even an occasional mention in *The New York Times* or *The Wall Street Journal* if you have genuine information to contribute on a newsy topic?

Never ballyhoo yourself. Be low-key. And don't expect instant results. But discretely let it be known to the press that they're welcome to come to you for quotes in your field; a reputation for perceptive, accurate comments can greatly enhance your image.

SUMMING UP

Let your clients' needs—as determined both by market research and face-to-face contact—drive your company. Marketing isn't a frill. It's a powerful discipline that can help your company more effectively serve clients. Technical savvy by itself is useless. So build a good marketing department, especially if you haven't time for the job yourself, and encourage your entire company to help draw in new clients. And buttress marketing with promotion. Finally, when clients do show preliminary interest, keep the marketing engine on track by researching needs and solutions in even greater depth. A good market intelligence data base—my next topic—can be no small help.

Do's

1. Organize an informal network to expand client contacts and intelligence gathering without having to increase marketing resources. Find small- or medium-sized firms with complementary skills and qualifications. Agree to pass prospective solicita-

---

tion leads to each other with the aim of teaming on the bids. Also consider the same relationship with hardware and software manufacturers that usually don't bid as prime contractors.

2. Keep your marketing function in the mainstream of the company's business, management, and projects. Then marketing can help keep existing contracts going—and also woo new business.

3. Realize that everything that marketing does—from generating leads to closing a deal—is a corporate project. it's not unlike a client's project. Let marketing orchestrate and manage all those tasks.

4. Make sure that your marketers have some knowledge and experience in the high tech services areas they market. They must know technologies, techniques, end-user applications (such as accounting, transportation, or personnel management), and general business principles.

5. Instill a marketing spirit in all your employees, especially professionals. Everyone should be prepared to identify prospects and leads from professional and personal sources.

Don'ts

1. Don't let marketing drop the ball on any opportunity. Make sure that all opportunities are pursued until they are either won or definitely don't show promise.

2. Never make marketing an ancillary or subsidiary staff function. If you do, you'll deprive the marketing engine of sufficient fuel.

3. Don't treat marketing as a sales organization, or marketers as salespersons. In the high tech services business, marketing must serve as the front end of contracts and projects, determining some of the clients' solutions in order to capture the business. In the industrial product business, salesmen operate after a product is designed and developed.

4. Do not ignore the speed with which the marketplace changes. You'll lose market shares if you ignore new techniques and technologies. Do not get left behind in a niche.

5. You should never forget your rivals' marketing capabilities, organization, and strategies. If they can afford to assign specific marketers to a specific client organization, you'll have a tough road ahead.

CHAPTER 4

YOUR MARKET INTELLIGENCE DATABASE

Hank Phelps worked for an engineering services firm doing business at the U.S. Cost-Pasting Administration. He learned that a salesman from International Monolith had sold dozens of microcomputers to Boondoggles, a CPA department outside Phelps's turf. "Can you help us?" asked the information officer for Boondoggles. "We've got all those boxes and don't know what to do with them." Phelps did. He spent hours telling Boondoggles how to configure a local area net for 200 micros to jabber over, and the department used his advice to prepare a solicitation seeking bids for LAN-related services.

A month later the marketing director of Hank's firm found out by chance about the solicitation. But by then the Boondogglers were already talking to another company, which won the $500,000 contract.

"Why didn't you report this lead to me?" the marketing director asked Hank.

"I'm not a salesman," the techie replied.

"I thought it was obvious—that you'd report all leads."

Alas, it isn't so apparent. Some techies might take to marketing as naturally as Sergeant York took to World War I. Others need encouragement, lots of it. Don't neglect it. Hank Phelps might as well have been a soldier who failed to provide his commanders with vital military intelligence. His information would have been immediately useful. And his superiors could still have stashed it away in a market intelligence database for future reference as well.

This chapter will tell you:

- What a good database requires—how it should offer vital information on which you can act.
- Eight broad categories of information, such as "General Economic and Business Information."
- What critical analyses you should apply to the database's contents.
- How you'll organize your database.

Follow the advice here and you'll be miles ahead of your competition. Many consumer products firms may systematically gather and file away facts on markets and rivals. But most services firms go about this more erratically.

Part of the problem is that you don't always know who the rival bidders are for specific jobs. Out of sight, out of mind. What's more, client contacts may be infrequent. Also, many services companies are happy with winning just 20 or 30 percent of the time when they respond to solicitations. They think they can't afford to collect, organize, and analyze data, at least not systematically. My reply is, "Phooey!" The big question is, "Can you afford *not* to do this?"

Your database, of course, won't just include information on rivals and buyers. It should also help give you a global perspective—that is, tell you what general business conditions are like, and how they'll affect your business.

With analyses using a good database, you can learn:

- How well the economy is doing. And you can assess its effect on your current and future business.
- How well your industry is doing, particularly among different market segments, and what the effect will be on your current and future business.
- How well your own company is doing. That includes feedback from clients, with indications of your strengths and weaknesses.
- How well your clients are doing, and what may increase or decrease opportunities for more business from them.
- How well your competitors are doing—their strengths and weaknesses.

You should then be able to draw certain conclusions that will tell you how to convince clients to do business with you, how to outmaneuver and outsmart your competitors, and what the most efficient paths are to business stability and growth.

WHAT A GOOD DATABASE REQUIRES

The foremost requirement for a good database is the inclusion of vital, factual information on which you can act—with confidence. Never bet your company on speculation or rumors disguised as "gut feelings."

I love intuition and will never ignore it. But I must earn my gut feelings; I must do my legwork; I am no different from a newspaper reporter, except that if I'm wrong, my mistakes will probably be much more costly. Systematic marketing intelligence in high tech services should help even more than in the consumer industries. Why? Because your market isn't as well defined as the consumer ones are.

If you're selling jeans you might aim your commercials at teenyboppers between 14 and 18 who want low prices and medium quality and maybe an endorsement from a six-foot, slim-hipped actress. America has blessed its jeans makers with millions of steady customers out there, all packaged in market segments like that one.

But in high tech services? Here you're going after a fragmented market that defies statistical prediction. Many clients come out of the woodwork at unexpected times. The federal government develops multiyear plans, but more often than not, it delays solicitations a year or two or even cancels them when priorities change in Congress or within the executive branch. Poof! Your market predictions and revenue plans go up in smoke. What's more, some agencies don't plan ahead and may buy within 30 days of telling you they want bids. You just can't take anything for granted, good or bad.

Say, the federal government wants to install Russian-proof computers—that don't spill secrets via radio waves—in hundreds of offices. You think bureaucrats will be pounding on your door for your valuable services. What a splendid possibility for the custom solutions at which you excel! But then you find that almost all the affected offices are ignoring the new directives, that the supposed mass market is actually a fragmented one. This, in fact, is what actually happened. So instead of thinking huge markets, you should look for chances to pursue individual opportunities.

You need a database capable of giving you factual information to take advantage of one-of-a-kind opportunities where the client is planning a purchase of services.

At the same time the database should help you respond to such opportunities—offering you helpful information such as facts on the specs and reliability of suppliers' equipment.

I'm not using *database* just in the computer sense; for example, supplier information might be on microfiche or optical disks. Either way, you could store thousands of pages in a small corner of a room.

Whatever the medium, your database system needs:

- Logical organization.
- Focus.
- Identification of sources on which to rely regularly.
- A way to analyze data and apply it to your company.
- Protection of data.

Logical Organization

Most services firms rely on sparse files gathered randomly and on staffers' memories. What ineptness! No general worthy of his stars would enter a battle without a good, centralized collection of the facts on which he could act. He'd want to know about his own resources and the enemy's.

And now I'll ask why services firms can't act likewise. After all, their battles last longer.

A simple way to organize files would be by "service function/ applications" and then by "market segment/niche." For instance: "1. Office Automation—Commercial Banks, Northeast; 2. Office Automation—U.S. Air Force Hdq." And so on.

Also under "service function/applications," you could include information from eight categories: (1) general economic and business information on markets to consider wooing, (2) what the market is buying and needs now, (3) government and public pressure, (4) information on suppliers, (5) new technologies, (6) facts on clients, (7) your firm's technical and financial standing, and (8) facts on competitors, including their market shares.

Even with this simple organizational style you might find you'd need a computer. A small company might start with an IBM clone with at least 20–40 megabytes of storage space (around 25,000 double-spaced pages) and a relational database and an electronic spreadsheet. Be prepared to perform "what-if" exercises. For example: If a supplier jacks up prices 10 percent, how will this affect your profits? And how much will it reduce your sales, given that a certain percentage of your clients won't buy equipment over a certain price? If a professional specialty is in short supply in your city, how will this affect salary schedules and control costs? I won't go into more computer-related details here since this is not a computer guide. Don't buy, however, without finding out much, much more, through books and courses that will give you the background you need to decipher the jargon in ads and in computer manuals. Especially learn to appreciate the importance of making frequent "backups," electronic carbon copies, as you add information to your database. Otherwise, if your computer fails, your business might die with it.

Focus

Focus the database around the nature of your business.

File away anything that might pertain not only to your company directly but also to your industry—including information about your rivals. But don't squirrel away peripherally related material that you can track down through a public library or (if you can afford them) electronic information services such as those of Dialog, NewsNet, and Dow Jones.

Source Identification

Where's the information to come from? You won't just be depending on your Phelpses. You'll systematically collect printed data as well. Here's a list of information sources that identify potential contracting opportunities, market trends, demands, competitors, and so on:

• The "bids and proposals" section of your daily newspaper's classified advertising pages, where you'll usually find local government needs.
• *Commerce Business Daily,* the federal government's source for contract seekers. It covers every type of product, from shoe laces to rockets—and every type of service from baseball umpiring to funeral directing to developing major telephone systems. You'll also find information on what companies have won awards, the amount of the award, and the buying agency. Available from the Superintendent of Documents, Washington, D.C. Telephone: 202/783–3238.
• The National Technical Information Center (NTIC), Cameron Station, Alexandria, Va., which contains a massive database on present and future technology, study reports, and so on. Telephone: 703/487–4600.
• The Defense Technical Information Center (DTIC), Falls Church, Va., a treasure trove for defense contractors. Among other things, you'll find the planned budgets and programs for all the military services and results of military supported research and development programs. You can also gain access to this database through the Tri-Service Information Center, which also stores detailed technical and management information on all current and future defense acquisition programs. The Tri-Service Center is at the Army Material Command, Alexandria, Va. Telephone: 703/545–6700.
• *The Federal Register,* which contains all proposed federal regulations for all government agencies. By reviewing proposed regulations you can determine prospects for new services and products and systems needed to support compliance to the regulations, along with requirements to modify existing products and services. The regulations affect state and local governments, too. Available through the Superintendent of Documents, Washington, D.C. Telephone: 202/783–3238.
• Other advanced program plans and budgets for federal and state programs. Get copies of them from principal contracting

officers and purchasing agents at each government site. In some cases the Government Printing Office (GPO), Washington, D.C. publishes periodic multiyear plans for civil agencies. These plans, such as the five-year plan for the Department of Justice, are identified in the GPO's published documents list and can be obtained on a subscription basis. Telephone: 202/783–3238.

• Employment advertisements in the daily newspaper. They're full of gems, such as tips on which companies are staffing, and the qualifications of the people they're seeking. If you're keeping up with outstanding solicitations, this will let you infer who is either planning to bid a job or who has won one—maybe you can approach the winner for a subcontract. These ads will also tell you about trends in technology requirements, and what companies are involved in each area of technology.

• Literature from industry trade shows and professional conferences. It will be a primary source of information on competitors and suppliers. So will your observations and conversations at the shows themselves.

• Professional and industry associations and their publications, indicating possible solicitations, subcontracting opportunities and employment advertisements.

• The Small Business Administration's PASS (Procurement Automated Source System) listing of small businesses and their capabilities for contracting and subcontracting opportunities. Obtain the PASS listing from Regional SBA offices. Main SBA phone number: 202/634–6197.

• Economic reports from banks, industry and professional associations, and Chamber of Commerce and trade publications. Their information will help you develop strategic market plans.

• Subscription information services such as Frost & Sullivan (New York), International Data Corp. (Framingham, Mass), and Dataquest (San Jose, Cal.). Those and other companies provide economic, development, market, and technology reports on specific market segments, related products, and services (such in reports from DATAPRO, Delran, N.J., 08075, and Auerbach Reports in Pennsauken, N.J., 08109).

• Client source materials such as organization charts and telephone directories. They're especially vital. Learn who reports to whom, who has what job, and the telephone numbers. Of all the sources, the charts and directories are the ones I use the most.

Two books packed with lists of contacts in federal, state, and local governments are:

- *How to Sell Computer Services to Government Agencies,* by Herman Holtz (published in 1985 by Chapman and Hall, 29 West 35th St., New York, N.Y. 10001).
- *The Defense-Space Market: A How-To Guide for Small Business,* by Philip Speser (published in 1985 by Frost & Sullivan, 106 Fulton St., New York, N. Y. 10038).

A Way to Analyze Information

How will you analyze facts and apply them to your company? Just how will the information relate to your business plans—or those that the client in effect makes for you? Using raw, unanalyzed intelligence is like trying to spread shelled peanuts on your sandwich. You don't want unprocessed pea*nuts.* You need peanut *butter.*

Protection of Data

In the future your whole database could be on one computer disk. Ideally you'll not permit anyone but yourself (or another responsible person) to gain access to any sensitive computer file. You should install appropriate security software that not only will prevent unauthorized accesses, but also will provide an audit trail of who has accessed what data.

And you should also guard against paper documents falling into the wrong hands.

EIGHT DATA CATEGORIES

Here are eight categories of information that you'll use as grist for your analyses.[1] Most of the items in the different categories

---

[1] By necessity there is some overlap among categories.

will relate to each other if you use the right database software.[2] Say, your client needs computer systems and you're pondering whether to make or buy the hardware. Where should you be able to find the information for this "make or buy" decision?

You should be able to answer questions conveniently from such *seemingly* unrelated areas as:

- What's the condition of the American microchip industry?
- Is your company's credit rating high enough to obtain the financing you need to crank out computers for the client?
- Is a manufacturing plant available where you can assemble and test the equipment?
- What is the availability of each of the components? What about discounted prices by lot? Delivery schedules? Warranties?
- Is a product available that you can buy off-the-shelf with equal or better warranty and at a cost lower than if you make the equipment yourself?
- Is there a market for the equipment beyond the current customer, so you can buy the components in larger lots, more cheaply? If there is, what about the competition?

Those are just a few of the dozens of items of information that you should be able to plug in.

Now, eight general categories of information to include in the database:

Category #1: General Economic and Business Information

You'll want to know *general* economic and business facts, whether you're (a) deciding to serve clients in a particular industry or (b) going after a specific company's business.

Say, you're courting steel companies. You know the old

---

[2] Alas, I don't have space here for a lengthier explanation of market analysis by database management software and other computer programs. See *PC Magazine, PC World, PC Week* and other computer publications and books, as well as guides such as DataPro, for explanations and lists of software sources.

heavy industries are in trouble, and that could cut more than one way. It might mean less money for your services. Or more positively, steel's woes could make your prospective clients more willing to try mill automation and other forms of high tech. Such possibilities are only the beginning; you should make many more queries. For instance, you'll undoubtedly also want to know if you can find good suppliers of automation equipment and if *their* industry is healthy.

Here are other important questions relating to general economic and business conditions:

• What about the national economy as a whole, and how it will affect the industry of your possible client?
• What about the health of the industry you'll use as suppliers of goods or services to help you satisfy your clients?
• And what about such oddball factors as the dollar-yen exchange rate? What will it mean if you're depending on products from Japan? How will all this affect your ability to serve your client?

Category #2: What the Market Is Buying

This category addresses marketplace demands, those mysterious drivers of business growth and crashes. Figure out what a demand will be in a market segment. If you're among the first to meet the demand, you'll be rich. If you're in the second contingent, you'll enjoy just a small piece of the pie. If you ignore the demand, you could go out of business. As I'm writing this book, some popular demand items include microcomputers, local area networks, office automation gear, computer-assisted engineering design, ambulatory health care centers, airplanes for commuter airlines, and equipment for industrial and computer security. And I'm asking myself, "What services relate to those hot sellers?"

The real trick, of course, is to be at the throttle of the engine leading the trend. How can you be so lucky? Well, you can start by keeping yourself posted on major constraints or gaps or problems imposed by new technologies. Most of the trade literature exposes these issues everyday. Someone is bound to provide the solution(s). The local area network business grew out of cries for communications solutions, to link computers from different manufacturers.

Along with marketplace demands that drive business responses are the special needs of certain client segments. But let's suppose here that you intend breaking into a new client niche. You must learn:

• What are the big issues in the industry? Suppose the industry is commercial banking. What is it doing about electronic fund transfer? Will the industry buy authentication boards to be installed in personal computers to guarantee that transfer requests come from and go to the right people—and that no one has tampered with the amount of money sent over the wires? As it happens, the industry wants the boards. But they cost $1,000 each and my research suggests the bankers won't pay more than $200. So I won't knock on doors until I can use a cheaper technology.

• How can you help clients put new products to work to fill their needs? Suppose a hot new spreadsheet for microcomputers has just hit the market. Who is going to provide the tools and techniques to either adapt the software to client characteristics or vice versa? Can you see the owner-driver of a semitrailer using Lotus 1-2-3, or Harvard Super-Project Management to run his business? Any way to make such tools useful to the driver or train him appropriately?

• What are your big prospects' general needs? Don't think 5 years ahead, but rather 12 to 24 months.

• Which needs are your rivals not meeting, and of those, which could you profitably satisfy?

• Who's selling services and products to the industry already? Some you might want to team up with, or compete with—or buy.

• How is the client base segmented? By size? By the nature of the business? By technology?

Category #3: Government Factors and Public Pressure

In general, how about government, tax, and regulatory conditions—and public pressures?

• What bills are coming from Capitol Hill or from your city

council, county courthouse or state house? Keep up with this at the federal level by way of a low-cost, part-time lobbyist, or other people in your industry, or a trade association. If you have the people and the money, also consider using *The Federal Register,* electronic bill-tracking services, *The Congressional Record,* and newsletters and reference materials from such companies as the Congressional Information Service and the Bureau of National Affairs.

• Will new legislation lead to business?

• If so, from whom?

• Might federal legislation force the states and local governments to act?

• Could the laws lead them to hire engineering services firms to set up new information systems or offer other high tech services? Suppose the feds require the states to administer a national driving license exam and send to Washington a list of new drivers. The feds and the states might then need a new network hooked into a national clearinghouse.

• What about opportunities created through other regulations such as new safety requirements for private industry?

• On the negative side, might new laws instead lower your profits or diminish your ability to develop new products, either for yourself or for your client to market to others?

• How about the legislation's effect on your inventory, if you have any?

• What about the general public attitude toward your prospective clients?

• What about public pressure? If the newspapers are full of headlines calling for the cleanup of toxic waste, legislation may follow. As a matter of fact it did. The so-called Super Fund bill— requiring generators of waste to set aside millions for cleanup— was a bonanza for some engineering services company. Some won contracts as high as $75 million. Lesson: Don't just read the newspaper and throw it away—systematically track public pressure.

Category #4: Suppliers

Now zero in more closely on the fate of your prospective suppliers of products or services. Remember, you might be at their mercy. If they aren't reliable for you, then your own company may seem amateurish to *your* client. Here are questions that your database should answer:

• What can you learn about the sizes of your suppliers' inventories, their pricing strategies, their deliveries and their warranties? What about the possibility of strikes? Will the suppliers reliably deliver on time? What's the delivery schedule promised? Thirty days? Sixty? What are the specifications of the suppliers' equipment? Are they supplying your rivals as well, and if deluged with orders, whom will they favor? Are they also supplying your clients directly? Will your clients have enough motivation to deal with you rather than with supplier directly?
• Do suppliers offer training, maintenance, warranties? What about other terms and conditions?
• How much influence do the suppliers have in their industries? IBM often sets technical standards for computers in general; Yourdon, for structured design techniques.[3]
• How up-to-date are the suppliers' offerings?
• What about another commodity: people? Can you find a good supply to serve your client: I don't advocate treating humans like corn or soybeans, but the brutal truth is that *most* of the people who do the grunt work are interchangeable.

Category #5: Technology and Performance

What's state-of-the-art to your prospective clients? What's most up to date to the rest of the world? Your clients may be ahead or behind. Other issues:

• What are publications—professional and otherwise—saying about the technology now? Go through the usual professional indexes and those of agencies such as the National Aeronautics and Space Administration (NASA) and the Commerce Department's National Technical Information Service. Also use private

---

[3] A structured design technique is a well-defined step-by-step procedure used often in the design of large information systems. The same methods can help design hospitals, other buildings, and subways. Structured design—with its big cost control advantages—is the opposite of the seat-of-your-pants approach.

services like DataPro and Auerbach (its evaluations are more detailed than DataPro's, but both have their uses).

• What technologies do the manufacturers plan in the next two to five years? Technological development often occurs in five-year intervals. In high tech services, you can't plan too far ahead; but you can at least keep up with the latest marvels that are about to emerge from the labs. You must keep up because your clients will demand the latest to keep them competitive and increase productivity.

• How will the technology affect your business?

• What about the pricing of this technology? How could it influence what you propose to your client?

Category #6: Facts on Clients

Beyond the questions already posed, you'll answer:

• What are management styles like? In general, what kind of people prevail? Ivy Leaguers? Then you *might* want your Harvard or Yale grad to be the contact person if he's qualified for the task. (I say "might," since you can't always go by stereotypes.)

• What are corporate goals, long- and short-term?

• What are the lines of authority within the prospects' organizations?

• Who buys goods and services and signs contracts?

• What's the bargaining power of the buyers? Do they write the specs without help, or are they leaning heavily on you for advice—in which case you'll probably enjoy greater bargaining power than otherwise?

• Exactly what are the buying processes like?

• Who makes the decisions? It might not be the same as the people signing the contracts.

• Do the buyers believe in competitive bidding or in sole-source contracts?[4]

• Do the prospective clients have any regular schedules, any yearly deadlines, for deciding which projects should win approval?

---

[4] See page 135 for an explanation of sole source.

• Who are your prospects' present and past contractors?
• What are the prospects' exact needs? If you're after govern-ment business, you'll obviously want to file away or summarize key items from five-year plans.
• Do they pay their bills within 30 days?

Category #7: Your Firm's Qualifications, History, and Financial Standings

What about the following?

• Accomplishments, track record, strengths and weak-nesses, special techniques, and solutions offered.
• Qualifications of managerial, professional, and adminis-trative-support people.
• Facilities, locations and distribution of talent, and other resources, including laboratories.
• Your firm's financial health.
• The size and kind of outstanding debt.
• Income sources and income growth curves.
• The market value of the firm and its assets.
• Whether the cash flow is healthy.

Category #8: Your Competitors

How strong are your rivals compared to you? Consider:

• What can you find out about them through business con-ferences, employment interviewees from Brand X compa-nies, trade newspapers and magazines, professional publi-cations, newspapers, and electronic databases? Develop a between-the-lines kind of smartness. Don't just read re-written press releases in the newspapers to find out what your rivals are up to; follow the example of a rival of mine in the Washington, D.C. area. Every Sunday he buys newspapers from cities from Boston to San Francisco and studies the classifieds, trying to find out what kinds of people his competitors are hiring for what jobs. Imagine the edge this gives him in anticipating the other Snark players' bids.

- What market shares do competitors have?
- What contracts have they won or lost?
- How are their companies organized?
- Are they mom-and-pop outfits or subsidiaries of GE or Boeing?
- Are they growing, and how fast?
- How do your prospective clients feel about your rivals?
- Which of their markets would you like to penetrate?
- How healthy are they financially?
- What are the strengths of your competitors' staffs?
- Is their management in good shape?
- Are there too many competitors now?
- In the case of each competitor, what is the distribution of services provided?
- What is the scope of technical capabilities offered?
- What are your competitors' normal labor pricing structure and strategies?

WHAT CRITICAL ANALYSES YOU SHOULD DO

Once you've assembled your database, you should find it easier to:

- Identify opportunities and threats.
- Spot your strengths and weaknesses.

Identifying Opportunities and Threats

Opportunities and threats come in many forms, ranging from technical to legal to economic. Often the same circumstances can either hurt or help a market. (See Chapter 7, "To Bid or Not to Bid?" for more information on the pluses and minuses that can influence whether you pursue individual contracts.)

While sizing up your specific marketing opportunities and responses, don't think strategically. Think tactically. In industry and government, buyers of high tech services typically plan for the next 12 to 24 months. Yes, the interval could be longer for big projects, especially government-related projects. But planners wing it month by month more often than you think. Often planners and prognosticators change their minds, finding that they're wrong—a fact that doesn't elude some of the more savvy chief executive officers (CEOs). Witness a recent phenomenon: the widespread disillusionment with (and unemployment of) corporate economists. In high tech services, anyway, this is an era not of strategy but of tactics.

At the same time you can recognize certain long-range trends, such as deficiencies or shortcomings in existing services or products. Thirty years ago IBM wasn't providing for the software needs of all customers. Suddenly a new industry appeared. Today thousands of independent programmers—people and companies—are writing standard or customized software to fill the gap. Even today, however, unfilled niches abound. Just read the splenetic columns that computer writers often write on the deficiencies of the existing products and services.

You can also identify openings created through new technology. Suppose you already design accounting systems for hospitals, but then you run across news of an invention—"The Intelligent Hospital," with computers assigned to the laboratory, X-ray, and pharmacy departments. Congratulations. You can now venture into a new market.

Each technology has its own window of opportunity in each of its markets. It's spring 1987 as I write this, and I know that local area networks are the rage right now in the federal government, but they haven't really caught on in corporate America. But wait. There is another market possibility: construction companies. They are now erecting the office buildings into which corporations will move within the next year or so. And several construction companies and real estate firms might very well want to woo tenants by offering them "wired" buildings.

Opportunities created by law have their own windows. Hanson was in on the ground floor when those new Medicaid revisions went into effect. Try to do the same during the drafting of new laws and directives for your own industry.

In the cases of both technologically driven changes and legal ones, the lines between opportunities and threats can blur.

Suppose you were helping to design or write programs for dedicated word processors in the early 1980s. Then microcompu-

ters hit the market. Before you knew it, corporations were forsaking the dedicated machines for the more versatile micros;
they not only could process words but also could easily run electronic spreadsheets and database and communications programs.[5] You had a choice. You could (1) pursue the ineffectual
course of running ads and otherwise trying to persuade corporate America to stick to dedicated machines, (2) leave the word
processing business, or (3) become involved in word processing
for microcomputers.

Of course, some threats seem just that: unmitigated threats,
such as a federal proposal to put a cap on the already low profits
of defense contractors in professional services.[6] Or what about
other menaces? A modified Federal Communications Commission (FCC) regulation, covering the certification of electronic
equipment to protect against radio interference, might delay
your entry into the lucrative systems integration market. Some
risks may even be great enough to cause you not just to forsake a
market but also the services business. But wait. If too many
firms leave, of course, then opportunities may abound for the
companies that tough it out. So plan to deal with changed regulations in time to prevent losses! In fact, see if you can't turn
threats into opportunities.

Spotting Your Strengths and Weaknesses

Now that you see a possible opportunity or threat, will your
company be capable of a strong response? And what are its
weaknesses? The appraisal tasks aren't always easy. You should
accompany each assessment of a strength with a suggested program to exploit it to improve the business. Accompany each
weakness with suggestions to counter it.

Consider your strengths this way:

---

[5] The word processing situation also shows how changing consumer preferences can be
an opportunity or threat. Still, in high tech, the technology itself helps create such shifts.
[6] I'm talking not about absolute dollars but about profit as a percentage of gross
contract sales.

- Your performance as compared to competitors'. If your track record is better in several areas, you must exploit these capabilities and demonstrate them to prospective clients.
- The ability to offer the right solution at the right price at the right time. Do you have this ability? Then rely on it for expansion.
- Your past experience. If your market share is growing, then investment could be a good way to jack it up further—and increase your return on investment.
- Your ability to provide the solution at the right price.
- Staffing. Do you have enough people? Are their credentials good?
- Geographical proximity to the client.
- Your past experience.
- The appropriateness of your problem-solving methodologies.
- The quality of your management.
- Your infrastructure. Can your legal department pick out important threats (such as penalty clauses) in contracts? And can your accounting department bill the client every 30 days so you enjoy a good cash flow?
- The dependability and reliability of your suppliers of goods and services.

Now analyze your weaknesses in the above areas and decide if and how you can overcome them.

For instance, you might think:

• We have too many project managers with little experience in management. Let's start some management development programs. Or find replacements.

• We lost too many bids because of price. We must find ways to lower our costs. Maybe we'll organize a low overhead shop.

• We don't have enough qualified staff to write proposals and bid on new jobs. Let's crank up a strong recruiting campaign.

• We're missing an opportunity to penetrate a rapidly expanding market. Let's either buy a company with experience or team up with one when we break into that market.

• Our unique problem-solving methodology has not helped our

recognition. Or won us any contracts. So let's start a strong marketing program. Simultaneously we'll test its efficiency, to find flaws.

HOW TO ORGANIZE A MARKET INTELLIGENCE DATABASE

You know the categories of information you're seeking and how you'll analyze it. Nice going. You're ready to organize your database and prepare people to use it. Some may be mere tipsters reporting to people who make actual entries. Again, however, you should ask for contributions from anyone with client contact, especially people familiar with buyer's technical needs.

Ready to start? Here's how you'll organize your database:

Sources

Change never stops, in technology, legislation, politics, business, research. But no one can read everything. So why not use the "journal clubs" familiar to many in the medical field?

Assign each member of the professional and marketing staff to read and review a selected set of publications. Monthly. He'll summarize *key* material to be entered into the database. In addition, on an ad hoc basis, marketers will enter information from client contacts. Assign a marketing administrator to maintain the database. Obviously he can feed in facts not only from his department but also from technical people.

Major Database Parameters

Specify the kind of information you should collect.

General Drivers
Prepare to consider such general drivers as:

- The kinds of information you'll need to define your direction in the marketplace.
- What your constraints are.

- What your risks are—for instance, new technologies or regulations.
- What your opportunities are.

The general marketplace drivers also include such demands and trends as:

- Need(s) satisfied.
- Client base by segments.
- Technology and solution base.
- Application areas and functions.
- Major players (competitors and market share).
- Major suppliers.
- Financial (cost aspects).
- Constraints and risks.
- Timing.
- Company response plan.

Legislative and regulatory constraints must also be part of your consideration of general drivers, including:

- Purpose or problem(s) to be solved.
- Client base.
- Technology and solutions.
- Application areas and functions.
- Major players (competitors, regulatory agencies).
- Major suppliers.
- Financial (estimated cost of solutions).
- Constraints and risks.
- Timing and schedules.
- Company action plan and position (strengths and weaknesses).
- Required organizational responses to pursue.

Also look at client base formal program plans:

- Program purpose (each).
- Status of program(s).
- Budgets.
- Office of prime responsibility.
- Technology and solutions.

- Application areas and functions.
- Existing contractors.
- Potential competitors.
- Potential suppliers.
- Constraints and risks.
- Company action plan and position (strengths and weaknesses).
- Sales goals (percent of client-base budget).
- Required organizational responses for pursuit.

Primary Drivers

Primary drivers include high probability active leads and prospects for each project, such as:

- Client (including relations with client).
- Requirements and solutions.
- Application areas and functions.
- Client budget.
- Sales value for life of contract(s).
- Booking value.
- Price strategy and win price.
- Life of contract(s).
- Expected revenue by quarter.
- Competitors.
- Technical and management strategies.
- Company strengths and weaknesses.
- Proposed organization responses.
- Win/capture strategy (for proposal).
- Action plan (costs/staffing).
- Status of pursuit and timing.

Low probability targets of opportunity are other primary drivers to be considered.

See active leads above for information items.

Strategic Support

- Company strategic market plans.
- Company plans of action and capture (past and present).

- Company qualifications and capabilities.
- Staff qualifications and capabilities.
- Summaries of current and past projects.
- Cost history of all projects.
- Cost history of marketing programs.
- Past proposals including costs and staffing.
- History of wins and losses.
- Past contracts including sales value and revenues.
- Competitors by technology and client base.
- Current and past client base.
- Applicable technologies.
- Supplier base.
- Information sources and purpose.

SUMMING UP

The database described here will need time and resources to build and maintain. But there's no real substitute. Every event recorded is an opportunity; every point of strength or weakness may be an important bargaining or balancing point. Realize that even seemingly trivial events can cause contract awards and business.

Realize, too, that for your company, for your new and current managers, a database can provide a legacy—they needn't start from scratch.

Properly established through the procedures I've described here, the database can offer most of the ammunition to:

- Make management decisions.
- Drive the course of the company.
- Help determine the extent of the company's success or failure in various activities.

And as you'll learn next, yes, you *can* use intelligence against rivals.

Do's

1. Use your marketing database to determine your plan of attack.

2. Keep your database up to date. Situations, players, laws—everything changes rapidly. Outdated data are no data.

3. Use the database as a company's legacy—as a history of its successes and failures, as a record of where you've been and a forecast of where you're going.

4. Let it help you track your position in the marketplace. Are you going up, down, or staying in the same place?

5. Use your database to keep abreast of your strengths, weaknesses, opportunities, and threats.

Don'ts

1. Don't shrug off any bit of relevant data as trivial. Each fact is a building block in intelligence gathering. If a purchasing agent gets a haircut that's not relevant; if he's sick, pay attention! He might be out on bid evaluation day.

2. Do not let your competitors see your database and have a chance to react to its contents.

3. Avoid depending *just* on intuition to make bet-your-company decisions.

4. Don't expect to thrive without qualifying your leads. That is, find out all you can about both rivals and prospective clients (including their programs).

5. Do not be casual about organizing your database. What good is information that you can't find and use? Be able to spot interrelationships.

CHAPTER 5

COMPETITORS

Tom Finch, a program manager at Sort, Inc., a professional services firm, spent several years helping the Department of Defense map out a computer network linked by satellites and microwave.

His net could have saved hundreds of millions of dollars annually—by improving Defense's capability to make travel reservations.[1] And over the years it might have meant $100 million in business for Sort. This was an antelope the size of 10 elephants, a lion's feast.

Finch, a true lion, defended his turf fiercely. He lobbied at the Pentagon and on Capitol Hill for backing and funding, toiled nights and weekends, slaved over the complicated technical plans at no cost to the government, and added up thousands of numbers to include in the budget. His expected buyers rooted for him. Finch's triumph would be their own. In the end, though, Sort lost out on the contract, and the Pentagon exiled Tom Finch's favorite colonel to Alaska. Too late Finch learned that a powerful competitor had bushwhacked him.

The whole sorry episode shows the need never to forget about the existence of hidden rivals. Finch had guarded his flank well among the colonels and midlevel managers at the Pentagon, but he left himself partly exposed among the gen-

---

[1] A transportation company might list its own carriers first and de-emphasize those of competitors with cheaper fares and more direct routes.

erals, and perhaps also at the very top of the Hill. Street-smart though he was, Finch didn't learn soon enough of the extent of the competitor's efforts against him. And so, at least here, he failed as a zero defect marketer.

In this chapter, I'll tell how to help avoid Finch's mistake. You must:

1. *At least try to find out about those hidden lions leaping out at the antelope from behind a bush.* In high tech services, you don't know who all your competitors are; you probably won't even know who the rivals are for a specific contract—not until the results are in. Sooner or later, however, you should learn what your rivals are doing right to win bids. And in other ways, too, you must aggressively gather intelligence on your rivals.

2. *Appreciate the different styles and ethics of competitors.* Here you'll find tactics to use for or against every common kind of pack—from The Decommissioned Garrison (staffed by ex-military men or other alumni of a client agency or company) to Superpowers (giants like GM).

Interestingly, since the buyer is constantly looking for a reason to reject you, he himself is an adversary—a point I'll elaborate on in the next chapter. You must consider the adversarial impact of competitors and buyers at once. Rivals may persuade buyers' bosses to steer business away from you—look what happened to Finch—or they may argue against you directly with the people with whom you're dealing. So think of the buyer and rivals together. Right now, however, for clarity's sake, I'll focus just on your competitors.

WILL YOU KNOW THE OTHER LIONS?

You're not in a footrace where you can observe your progress versus your rivals and watch them gain or stumble. Beware of the opposing lions that you can't see.

A few of the smarter animals play a good Snark game. They can lay down winning cards, or leap out at you from behind a tree, when you least expect them. You might not even know they're foes on a particular bid. And so one of your most important tasks is to try to identify your hidden competitors. Clients—government or private—normally won't disclose a winning contractor's offer.

And so the success formulas of your rivals remain outside your scrutiny.

These other lions fall into two categories:

- Those with which you're familiar. For instance, you might know of just eight companies in the United States qualified for the work.
- Spoiler lions that leap out of the bush: the mysterious and troublesome ninth companies.

What's more, the mix changes constantly. Next time only five of the original nine lions may be scrambling after the antelope. And there may be two absolutely new rivals.

How Come the Other Lion Caught the Antelope?

"Why'd I lose?" many in Finch's place would have asked.

An even better question, however, would have been: "Why'd the competition win out? Why did the trade association get the contract—even though this project had been my baby from almost Day One?"

In this case the exact answers seem to have wafted away with the cigar fumes in Washington's proverbial smoke-filled rooms. Finch doesn't know who pulled which strings.

But the same principle—the need to find out the reason for success—would apply elsewhere.

Figure out why the competition got the contract, not just why you lost. Don't content yourself with the buyer's explanation.

For instance, if the buyer says your rival proposed a better-qualified staff, you should ask yourself:

"How'd they manage to line up the right people at the right time to satisfy the buyer's needs? What did they know that I didn't? And can I replicate their success next time around? Just how can I make the buyer like my proposals more?"

Observe closely and you'll find yourself not leaving competition to chance. To win you must control your game.

And if your adversary has learned some of the rules better than you have, then you'd better catch up.

Getting the Lowdown on the Other Lions

Can you find out what cards the other lions are holding in the Snark game? Or even who they are? Those lions so often wear masks when they play cards. How to ferret out those hidden competitors? And how do you learn the strengths and weaknesses of known rivals and anticipate their moves?

Only rarely do you know whom you're bidding against. Yes, XYZ Corporation might usually bid on this kind of communications or maintenance solicitation. But you never know for sure—or even how many other lions will lay their cards down.

For instance, on many government bids, 100 or so firms will ask for solicitation documents such as a Request for Proposal. Perhaps 50 will attend a prebid conference to learn who the other lions might be.

But in the end? Fewer than half a dozen might actually bid.

Your mission, should you accept it, is to try to puzzle out such information as:

1. *Your rivals' prices.* You're not Wendy's competing with McDonald's. You haven't any idea what all your rivals' prices will be. And yet you'd like to know. If only you could raise your price without exceeding the competition's! At the least, you don't want to go broke to undercut their bids.

2. *Their solutions.* Will the winning solution involve state-of-the-art techniques? Will the solution be a customized one or come off the shelf or be directly transferred from another contract—thereby reducing costs and reliability problems? Can your rival cheaply upgrade the hardware, software, whatever, and avoid paying for replacements?

3. *Their techniques—both in pursuing business and in carrying out the proposed work.* How will you fight back? You're trying to outguess the competition. They're trying to outguess you. In fact, you're all playing my favorite Sherlock Holmes-

Professor Moriarity game—guessing what they're guessing about what your're guessing.

And then one day you learn who caught the antelope. "Who? Who?" That's what you're thinking when the buyer—the antelope breeder—utters a name of a lion who supposedly wouldn't ever stalk through your part of the bush.

The next month you bid again on another contract. Once more a strange lion's name pops up in the end. This should tell you to alter your competitive strategy on each bid, as you can expect a different mix of competitors again.

Here, however, are some ways to track your competition better.

1. *Watch the classifieds in local and national newspapers and in trade journals.* Say, IBM is running a giant advertisement for fiber-optics specialists familiar with local area networks. If you're working in AT&T's office automation branch, then a big blue lion might be breathing down your neck—since newspaper ads can hint of future employment and marketing plans.

2. *See if your competitor is active in professional organizations serving clients.* Suppose the marketing director of a defense contractor is an officer in a retired officers association. Then you can almost take it for granted that he's making business contacts there—including perhaps those with the buyer's organization.

3. *Learn how strong and how weak your rivals' people are in various specialties.* Find out through professional organizations. If you know names of your competitors' key people, see if you can track down articles by them—through the indexes of professional journals. Go to conventions. Attend presentations. If you feel comfortable with this, place your own want ads and interview people from your rivals who respond. Pick their brains. Who knows, you may actually end up hiring them—except that if you do, you'd better make certain they'll keep secrets better in the future.

4. *Ask one or more of your rivals to team with you on an upcoming job.* Request briefings on their companies and capabilities. Learn their strengths and weaknesses. Find out what special methods they use and the compositions of their professional staffs.

HOW TO FIGHT (OR CHAMPION) THE DECOMMISSIONED GARRISON, THE MARAUDERS, AND OTHER KINDS OF ORGANIZATIONS

Your competitors' styles and ethics will vary, and you must respond accordingly.

Here is what you need to know about "The Decommissioned Garrison" (staffed by ex-employees of your prospective client), "The Marauders" (staff piraters), "The Fallen Knight or Lost Brigade" (the techie-run firm), "The Adoption Agency" (the client stroker), "The Turfers or Territorials" (keepers of contracts or market segments), "The Mendelian Crossbreeders" (acquirers of other companies), and "The Superpowers" (the IBMs).

The Decommissioned Garrison

Many Defense contractors do employ ex-service people. And why shouldn't they? Every company has the chance to hire service alumni, and the criteria for choosing them aren't capricious. They must show technical or management skills. In short, they have to be of value to their employers, and that's usually the case. Who's better suited to work as a contractor's quality assurance engineer—looking for flaws in aircraft parts—than an ex-pilot trained in engineering as many of them are? Or what about the perfect candidate for a job as a contractor meteorologist for the merchant marine? Couldn't he have done similar work for the military?

More closely related to marketing, how about this example? The chief of a Navy computing facility might go to work for a services firm specializing in the management of such centers. He might know what contracts are going to be issued by the center. In essence he's a commodity in the marketplace, and many contractors could bid for his body.

The same old-soldier principle, moreover, would apply to

services companies outside the defense industry, including those catering just to other businesses:

- A health planner with the Department of Health and Human Services might end up at a major hospital chain seeking federal health funds.
- A Department of Transportation expert may go to work for a major builder of interstate highways.
- A purchasing agent for General Motors might retire to work for one of GM's major suppliers, where he could use his experience and contacts to win and manage subcontracts.
- A young engineer-manager may leave AT&T to work for a firm interested in the phone company's strategic plans for its long-distance service.

What's more, there is another kind of decommissioned officer—someone hired away from a competitor to gain the same kind of information or contacts that client alumni can bring. Many of the same principles applying to the client alumni would apply to them.

Still, an alumnus isn't a sure meal ticket for his new employer. Often the usefulness of both the inside information and contacts will rapidly start to diminish. The government's plans may change, for instance. The old boys who made them in the first place may be transferred, retire or die (no murders, just normal statistical probabilities). And rival contractors may hire their own Decommissioned Garrisons. Anyone can.

So the original contacts may not necessarily lead to new contracts. And there are other negatives.

The old generals or civilian equivalents might have influenced projects budgeted at $50 or even $250 million. But usually they wield this clout only within particular programs.

What's more, the law may prevent old soldiers and others from contacting their old comrades directly to drum up new business—for at least two years after their departures from the military or government. And so they must limit themselves to behind-the-scenes advisory roles.

That's only right. It would be outrageous, however, for new laws to forbid the retirees' participation *entirely*. The public often benefits when old military and civil service hands can use their knowledge of government policies and procedures to help employers come up with more responsive bids. If contracting firms aren't doing their marketing homework, this is one way to help the government receive responsive bids.

Here are some typical examples of public benefit:

- A retired general might help a defense contractor come up with a more useful weapon than would someone who isn't so familiar with the Pentagon's needs.
- A former FBI man may use his government know-how to help plug up security leaks.
- Working for a consulting firm, an alumnus of the Occupational Health and Safety Administration might help develop a better way to measure toxic fumes in chemical plants. Perhaps bureaucracy prevented the ex-OSHA man from doing this in his incarnation as a government employee.

Similarly, a former marketing executive with General Electric may go to work for a consulting firm serving GE—and may come up with a better way for the electronics company to fight off Japanese imports.

And so a flat ban on Decommissioned Garrisons—public or private—would only cripple America. Nothing can replace the judgment and maturity that grizzled old-timers can bring; even the whiz kids in *The Soul of a New Machine* needed middle-aged people to point them in the right direction. New technology isn't enough.

Probably, in fact, more than 90 percent of the companies in the professional services and engineering services business have tapped the skills and insights of ex-insiders.

If You're Fighting a Decommissioned Garrison

Must you wave the white flag the moment you learn a Decommissioned Garrison is competing against you? Absolutely not!

Your tactics and strategies should be to:

1. *Find out if it would be useful to post your own Decommissioned Garrison.* Visit the offices of the prospective client. See if any former employees are now working for your competitors. If so, prepare to play by their rules. You, too, should play the game. You have equal opportunities to hire the same old boys.

2. *Hire only the old boys who can do you the most good.* That's how Sling Defense Systems outwitted Philistine Technology when they were both fighting over a contract for programming, security, personnel and accounting services—a multimillion-dollar plum. Philistine hired two recently retired colonels to help crank out the proposal. They had written the specification and statement of work for the solicitation themselves. But Sling Defense Systems won anyway. How? It hired the general who had been the colonels' boss. Another subordinate still worked for the prospective client, and he successfully lobbied for Sling.

But don't just consider past and present influence. Also keep in mind:

- Academic credentials.
- Technical and professional qualifications.
- Understanding of the kinds of plans and programs that the prospective client has in mind
- The ability to help you in areas outside their old spheres of influence.

Hire a Garrison member only for a particular job if he lacks the skills and contacts for other projects.

And beware of taking on people who have put in their 20 years and now are looking for a place to retire. You're not Miami Beach. Loyal old-timers, grateful for their jobs, are fine. Just be sure that all your people—young and old—show energy and initiative and are quick-witted enough to learn new tricks.

Problems not withstanding, Garrison members' advice can be invaluable—even if the law may temporarily forbid them from speaking, writing or talking to their former peers or subordinates.

3. *Market vigorously.* Show you can understand the client's needs even better than the ex-insiders.

4. *Propose the use of newer, better tools on the job.* Old-timers may be less open to the new. Turn this to your advantage. An engineering company, for instance, may fight the Decommissioned Garrison through the use of computer technology that the rival's people are too "mature" to want to understand.

If You Yourself Are in a Decommissioned Garrison

1. *Don't take your connections for granted.* As you've read, your rivals can develop those of their own.

2. *Do not commit ethical or legal breaches.* The public relations troubles and possible political complications aren't worth it.

3. *Make sure that newcomers from government or the military are quick to pick up competitive habits.* They must seek out business. It isn't enough to laze back, waiting for field people to bring in customers. Many bureaucrats and military people have received raps on the knuckles—or worse—for showing initiative. Now they should know that you'll give them another reaction: applause. Make them aware that they're no longer in government, where bosses may underemphasize initiative or creative techniques that bend traditional rules.

The Marauders

Beware of such packs!

Marauders will pirate your staff and steal your clients, ideas, and techniques. They use stealth and cunning. And often they act in the guise of a friendly prime contractor or subcontractor.

Then, when you're off guard, when you're like a trusting lion rolled over with his vulnerable stomach exposed, they'll attack you and steal your antelope.

> Let's say you need Thatch, Inc., as a subcontractor to provide the people with the right skills and credentials for a job since you don't have them in-house. You win the contract. But now Thatch's people can visit the client themselves. They discover how you do business and maybe even how you tote up your costs and prices. You've exposed your guts.
>
> What's more, Thatch will bad-mouth you or your people to your client.
>
> It might even place its own company name on your drawings and other important documents and make job offers to your managers and staff.
>
> You're furious. But there's only so much you can do. Thatch, Inc., is thinking of offering you a subcontract on another project

and you don't want to jeopardize it. You know it's all legitimate, that Thatch really needs you. So a lawsuit is out of the question here; in fact, with only so many resources even to handle normal business, you're not about to open up this kind of a front.

Another trick might be for the Marauder to call up your company and ask you to be a subcontractor.

> You're flattered at first. Slee Z Systems is one of the largest professional services firms in the country, and here's the vice president on line one saying: "We're very interested in giving you a subcontract worth tens of millions of dollars. But first you've got to show you're really qualified. It's a price-sensitive job. We need to see your books."
>
> With a $10 million-plus carrot dangled before your eyes, you indeed open the books up—showing the payroll and chart of accounts.[2] Weeks pass. You learn that Slee Z decided not to subcontract.
>
> A few months later you're bidding against Slee Z. Guess who wins?
>
> Unfortunately, this example isn't hypothetical. It happens often. So do the other tricks that the section below will help you guard against.

If You're Fighting the Marauders

1. *Go ahead and team up with a Marauder on a project or participate in other deals, but insist on legal agreements to protect yourself.* The agreements might specify:

- That there be no hiring of staff from each other for at least a year after completion of the job. If possible, try to avoid supplying resume information about your staffers until the Marauder signs an agreement. He could use it not only to pirate your people but also to help learn who your clients are.
- That you'll disclose price information only when absolutely necessary.
- That if you do disclose price information, it won't be in the Marauder's permanent database.

---

[2] A chart of accounts is a list of all items—from stamps to floppy disks—that you regularly pay for.

- That your partner won't market your client.
- That all client contacts be made only through your office, with your own managers present.

Mind you, such agreements aren't foolproof. Your friendly Marauder might call on your top technician at home, for instance, asking him to send a resume. You may end up thinking that he himself initiated the resultant job switch.

A Marauder might also use as simple a ploy as inviting your client to a Christmas party.

Alas, even the most carefully worded agreements against marauding may be just a form of damage control, as opposed to true prevention. Consider this before undertaking the risks of dealings with outfits like Slee Z.

2. *When placing employment ads, try not to specify the name of your company—if possible.* The Marauders might still guess. But at least you've somewhat lessened the chance of their finding out what you're up to. Of course if you do place a blind ad, you most likely will discourage some prospects from applying. You also might experience the frustrations of seeing some of your own people apply for employment (ostensibly) elsewhere. But those are the breaks if you're working hard to thwart the Marauders.

3. *Be wary when you're interviewing people answering employment ads, even blind ones.* They just might be spies from your enterprising competition. Don't run off at the mouth and leak too much information during employment interviews to entice a hotshot to work for you. The red flag should go up if the interviewee asks probing questions about sensitive activities unrelated to the job sought. Beware if he's too nosy about pricing practices, names of present clients (as opposed to names of past client companies), when certain contracts are ending, names of principal professionals, and so on.

There are common-sense exceptions. Obviously something would be wrong with a prospective marketing man if he weren't curious about your main lines of business, about what kind of clients you sell to, and about your marketing procedures and reporting mechanisms. But beware of *too much* interest in peripheral matters—don't spill the beans on how you'll win your next big contact.

4. *Maintain good relationships with your clients.* That way, they're more likely to come to you directly rather than through the marauders you might employ as subcontractors. And simple loyalty can make the clients think twice about switching. Retired Captain B. Beard was a first-class manager working for Pangloss Research, but planned to start his own company. He pestered a Pangloss client to give him a contract to make this possible. Not only didn't the client comply, the organization even reported Beard to Pangloss—which he indeed left eventually to go off on his own.

5. *Try to avoid entanglements with Marauders that could lead to an unwinnable lawsuit, or one that at best would simply prove you were right.* Say, a Marauder promises to provide you part of the job—maybe one fifth of the labor hours or of the dollars. It's possible that an unscrupulous Marauder will either (*a*) not give you the contract at all or (*b*) stint on the amount of business thrown your way. Be familiar with the representatives of the Marauders with whom you're dealing. Perhaps sometimes the integrity of individuals can transcend the integrity—or lack of it—of the company. But normally that isn't the case. Most people either adjust to the informal ethical codes around them or move on to another firm.

Never forget the typical Marauder's recipe for growth: crushing the other guy's backbone.

If You're Working for a Marauder

1. *Be a Gung-Ho Marauder.* Marauding is part of the normal professional services environment. It's in the air. A services industry without marauding would be like San Francisco without fog. So be both cunning and cautious. Make sure nonraiding agreements aren't too limiting or constraining. For example, be certain that your prime contractor's agreement allows for some leeway in communicating with its client on, say, contract technical matters. You might inform your prime contractor of the times that you do converse with its client during business hours. That's one way to satisfy the main contractor that you're living up to the agreement. At the same time, by superior work and perhaps subtle hints within the limits of fair play, you can let the client appreciate the advantages of dealing directly with a company as technically accomplished as yours is.

2. *Gather intelligence aggressively.* Go ahead! Call Pangloss Research for that marketing job interview, so you can find out about clients. Ask a colleague to phone about work as an engineer so you can find out how much your competition pays its people—no small part of project costs! Better still, ask third parties, not connected to your firm, to seek a Pangloss interview.

How ironic that most of the services industry doesn't know how to compete—but takes such James Bondery for granted.

The Fallen Knight or Lost Brigade

Lancelot Technology[3] starts when the Knight's old company fails, or when he tires of arising at 5:45 A.M. everyday to be able to find a vacant parking space near the door at Camelot Systems. What's wrong with those people? Why isn't their best techie good enough for a reserved space inside the moat?

The Knight coaxes a small but lucrative contract out of a friend of a friend. And perhaps a venture capitalist, aware of our techie's formidable credentials, throws in several hundred thousand dollars.

At first the Knight's headquarters are in his rec room; specifically, the ergonomic computer table between the Nautilus machine and the VCR. Then two buddies join him. Within weeks he's rented space in Reston, Va., or the Rt. 128 section of the Boston area, or Silicon Valley, or wherever Knights joust nowadays over Star Wars bones and subcontracts from General Dynamics.

However good-hearted, the Knight may be miserly in the worst techie way. Nor will the Knight fritter away money on a decent reference library or departments for technical editing or contracts processing or a lawyer who knows the business.

The Knight thinks he's on a roll. He and the old buddies from Camelot Systems—the Lost Brigade—steal away business from several other small-fry. The roll goes on. The Knight works 70-hour weeks. Little does he know that in the end he'll fail. A big reason? He keeps himself fully billable. The Knight not only

---

[3] I know, I know. Somewhere out there must be an actual company named Lancelot Technology. Rest assured that my own Lancelot is a fictitious name.

manages his contracts, he does technical work, so that he has little time put aside to run his company itself and no time to scare up new business. He prides himself as an armor-on executive.

What's more, the good-guy Knight grants himself and his buddies perks of the kind that he'd dreamed about. Meanwhile the clients keep saying they love Lancelot Tech and other contracts are in the mill. The Knight slackens off a little. He even finds time for a Jamaican fling with Guinevere.

Then, just after the Knight has hired a dozen new staffers, Defense undergoes some surprise budget cuts—vaporizing the Knight's three biggest contracts, as if they're enemy missiles.

His clients dig up some obscure clauses to worm their way out of their commitments. The Knight frantically searches for replacement business. It's storming outside Camelot. But there's no shelter, no protection. Without good business planning, the Knight lacks cash reserve.

And even more important, devoid of marketing skill, the Knight can't scrounge up business to replace the lost Defense contracts.

The Department of Health and Human Services may be doing some medical research through which the Knight's electronic wizardry could hasten a cure for cancer. But the Knight will never know. Defense—or more specifically just one program within Defense—is his territory. Nothing else. Ask him what HHS means and he'll say, "What's that? A British ship?"

If You're Fighting a Lost Knight

1. Have pity. Most Knights fail.

2. If you lack pity, underbid the Knight. Your cash reserves are bigger than his. Because he's so small, the potential contract is likely to be tiny, too—making it easy to eliminate this potential competitor. Just market his current client base; he himself lacks the time.

If You're a Knight or Are Jousting for One

1. *Make sure you have special prices, special technology, special tools, special skills, special personal qualifications, special anythings, with which to fight the competition. With which* company would you rather deal? One just starting up or an established firm?

2. *Bone up on business practices.* Take courses if need be. And talk to other entrepreneurs—those not competing with you—to find out what works and what doesn't. Learn that if you're managing a company, you don't just manage projects.

3. *Don't stint on cash reserves.* Even a company with one or two professionals should have a big kitty to tide people over during the first lean year or two. For two $50,000-a-year techies, that amounts to about $500,000, which will pay their salaries and other business costs in case no business comes their way. If they gross $500,000 for the two-year period, their net profit after taxes will be about $25,000. Now let's say that the backlog is just $75,000 of business for the third year. Then the reserve the Knights need to survive for that year is $150,000. The usual formula is to accumulate four times each person's salary in reserve if your backlog is minimal for the first year in business.

4. *Avoid the grasshopper syndrome.* Market ahead! Line up clients for wintery times! And work on expansion and renewal of existing contracts.

5. *Invest as soon as possible in auxiliary services such as a good reference library.* Typical high tech companies need a good, convenient library facility to support technical projects and, especially, to help keep their proposal development on schedule. You'll lose business since you can't find critical reference materials promptly—if at all—at a public library. Also, many clients call in for answers to technical and contractual questions. There is no time to drive to the Library of Congress for answers.

The Adoption Agency

You've probably seen those skeletal Third World babies in magazine ads that plaintively beg for your support. Well, charity isn't quite what the Adoption Agency has in mind. This eager bidder wants to baby someone else—the buyer that can support *it.*

That means everything, from fishing trips and free office space for the buyer to poker games where the bidder consistently and gracefully loses in hopes of winning a contract later on. The poker ante is low. Just one lost poker dollar, however, might mean several thousand won contract dollars. So goes the adoption agency's logic.

Through it all, the buyer is indeed flattered—but probably still won't let his daughter marry the contractor's son.

Don't think the marketing department isn't trying, however.

The marketing people reign supreme. An Adoption Agency is the opposite of the Fallen Knight's company, where marketers are greeted as warmly as hard disk failures.

The Knight's operation is one dangerous extreme. The Adoption Agency is another. The marketers pay too much attention to stroking the client—when very often this quickie is foredoomed. This approach isn't just handholding. You overwhelm the buyer with an office at your headquarters, for example; an indefinite loan of a new PC; prepaid trips to overseas conferences; or arrangement of low-interest loans. Warning: If you fail once to gratify the buyer, you're out. And even if the buyer is susceptible to this approach from Bidder A, he may be equally so to approaches from Bidder B; and besides, he's probably seeing through both A and B. He knows they're interested in more than the pleasure of his company—unless that's the kind of company with a large C, an Inc. after its name, and plenty of contracts to dole out. He might actually prefer that the bidders be more businesslike.

If You're Fighting an Adoption Agency

1. *Socialize but only within limits.* Your rivals may glad-hand the prospective client, but you can succeed by not ignoring matters of substance. This makes for better, more stable relationships between buyers and contractors.

2. *The people cultivating the buyer should share his interests if possible.* But at the same time they should be knowledgeable about the possible contract under discussion.

If You're Fighting for an Adoption Agency

1. *Realize that the best way to stoke the buyer is to bolster his bureaucratic standing by coming up with a good, responsive bid.* And why not also prove that your company is financially stable and boasts a good track record and people with the right credentials? Go ahead and play poker—and lose gracefully!—but while you do so, why not see if you two can work together to solve the buyer's problems?

2. *Don't overstroke.* To avoid feeling too indebted to you, some buyers may actually go out of their ways *not* to give your bid the benefit of the doubt. What's heavy stroking but a source of unwanted friction?

Observe Friedman's Poker Rule. Don't win more than $10 or lose more than $40. A client will probably think you're a hustler if you win more than $10, and pretty dumb if you lose more than $40. Factor in inflation. Base figures are for 1987.

Above all, people at Adoption Agencies should operate within the bounds of conscience and the law. Consult with your corporate attorney or even hire one privately if need be; do not blacken your image by continuing to work for an Adoption Agency if your boss leans on you to act questionably. Otherwise you later might have trouble finding work at a reputable services firm.

Turfers or Territorials

Retired General T. T. Cratchit grew up in a poor neighborhood. He clung fiercely to his few toys.

Decades later, General Cratchit founded a high tech services company. The old habits remained. But this time it wasn't toys. It was markets and client segments. No matter how low his competitors bid, General Cratchit was ready to outwit them and drop his prices lower and lower, lest he lose clients. It wasn't just a case of missing out on the business. This Turfer didn't want anyone else to have it.

Emulating his father's old boss, General Cratchit set up shops with low pay and low overhead in areas where living costs and rents were below norm—where engineers would toil in converted warehouses without windows or carpeting.

General Cratchit felt he couldn't compete in other niches. So he was always the low bidder.

And so far it's worked. The clients keep renewing and renewing.

Their contracts—extended with a minuscule marketing budget—may call for as many as 500 professions and offer over 1 million person-years of work. If the Egyptians build new pyramids, you can bet they'll hire Cratchit Technology.

If You're Fighting a Turfer or Territorial

1. *Count on spending a long time befriending the buyer.* See him at professional association meetings. Also drop in for casual chats and perhaps offer freebies such as feasibility studies at no charge to help him make decisions—for example, on what equipment he should buy for his next upgrade. Don't excessively glad-hand. Just make sure your face and generosity become familiar. Most likely, personal relationships are no small reason why he's been dealing with Gen. Cratchit for so long.

2. *Be prepared for less income return.* You're in it, as they say, "for the duration." So are some biggies—the Pan Americans, the Boeings, the TRW—who love to "turf-in" at places such as the Cape or other NASA facilities. The buyer pays less for overhead and administrative costs. But the Turfers know the long-range rewards. Eventually, just through sheer volume, they make up their initial losses.

3. *Consider setting up a new subsidiary with wages, overhead and other costs lower than your usual ones.* In fact, if you beat Cratchit Technology, you might keep most of the low-paid people assigned to the job you bid on. Only the employer's name on the paychecks might change.

4. *Try to woo the turfer's top managers.* See if you can't hire Cratchit's top managers at higher salaries, under the proviso that they also bring along some of the top techies. Cratchit will now have problems writing his rebid proposal.

In short, you have a wonderful boat-rocking opportunity! Bear in mind that outfits like Cratchit Tech have a substantial turnover rate. The General loves to use young, inexperienced people and retirees. The Turfer's massive operations may even set the wage rate in the client's area. Probably it's lower than average. And now you can win the turfer's people over—especially the younger ones—with your willingness to pay more.

If You're Fighting for a Turfer or Territorial

1. *Although you'll want to keep costs low, don't skimp on quality.* Get rid of any incompetents whom you've kept just in the interest of economy. Reward people who produce good work in abundance. Superior performers used to low pay will find average pay to be a wonderful surprise. That builds loyalty. When you do economize, consider doing so by way of low-rent offices rather than low-rent people. Don't jeopardize your reputation for quality work. Who knows, perhaps a surfeit of the Turfer mentality contributed to the space shuttle disaster.

2. *Try to diversify.* Why not milk some of the cash cows to finance the marketing campaign you need to acquire new business in other, more lucrative environments? It's too dangerous to rely on too few contracts and many competitors. Just look at all the ups and downs of programs such as the space shuttle and (very likely) Star Wars.

Mendelian Crossbreeders

For 20 years, you competed against Tantamount Research—a professional services company of the same size. Sometimes you win, sometimes Tantamount does. It's an equal match.

Then you read in the business section that Tantamount is now one of the many tentacles of Hydra Inc.

Hydra Inc.'s new Tantamount Division will enjoy more funding, more marketing experts, more resources of all kinds, and you wonder how you'll compete. You wonder how many Tantamounts are out there. Nowadays sales and mergers of corporations are common; just look how General Motors swallowed Electronic Data Systems Corp. But conflicts sometimes flare up. Managers with different styles—for instance, traditional corporate versus entrepreneurial—clash.

Buyouts and mergers between high tech services, however, are in some ways less of a problem. Aren't most of them clones of each other, anyway? Of course there still may be unpleasant differences. So the best Mendelian Crossbreeds buy or merge with firms with complementary clients and skills—while trying to filter out the weaknesses.

If You're Fighting a Mendelian Crossbreeder
1. *Consider buying or merging with another company your-self—or setting up consortiums or perhaps long-term joint ven-tures.* If possible, avoid mixing with Marauders. Try to go after companies with complementary skills and markets but similar values. See whether your goals will blend well with those of the other companies—and whether you'll be able to allocate power comfortably to the incoming executives. If you can't allocate, then you'd better stick to buyouts!

2. *If you want to grow larger, look for niches ignored by bigger rivals.* In recent years, the giants might have squeezed you out of big systems integration jobs. But look at the growing niches. Consider teleconferencing systems to link chemical com-panies' plants, communications security for banks, and artificial intelligence for financial planners and owners of home com-puters. And what about local area networks for new buildings—work for which construction firms can give you subcontracts?

If You're Fighting for a Mendelian Crossbreeder
1. *Take advantage of your deeper pockets to underbid the competition if you see a client that can send plenty of new busi-ness your way in the future.* Scare competitors. But do so within the limits of good (future) financial sense.

2. *Try to eliminate overlapping services. For instance, you can combine laboratory resources.* At the same time respect the individuality of people and research objectives. Don't merge pro-fessional services marketing with industrial product sales; each has a different style and method.

3. *When you acquire new companies, weed out the incompe-tents and other threats to your efficiency as soon as possible.* Protect your corporate culture from bad influences, especially in mid- and upper-management ranks. Some of the worst influ-ences might be from integrating services companies with those in products.

Superpowers

Now we face the real giants, the Bechtels, the IBMs, the huge architectural and engineering firms, and the AT&Ts and GMs.

Some of the biggies also fit the Mendelian Crossbreeder category.

But there's a difference. The giants often do not need to buy companies to break into the services market; they may find the skills in-house.

The new services divisions almost instantly find customers—among existing clients. Bolstered by the corporate equivalent of a most-favored nation policy, they relentlessly steal away the business the Superpower has given you in the past.

Take a Westinghouse power plant. You might have supplied professional engineering services and related studies in security, facilities design or communications. Perhaps Westinghouse considered the work to be too limited to worry about doing it in-house. But that's changed. Now its specialists eagerly vie for such contracts.

Most of all, of course, people from companies such as AT&T and IBM are after the gargantuan systems integration contracts from the U.S. government that are worth as much as $4 billion and last up to 20 years. The brighter side of this is the chance you have of becoming a subcontractor. Still, if you're small, you have a right to feel frustrated. The government isn't exactly rushing to dole out contracts that large to average-sized services firms, and money from the big deals can help finance the giants when they're competing against you. Size brings other advantages. The Bechtel-sized firms can buy thousands of square inches a year of institutional advertising in prestigious magazines and newspapers to build on their established reputations.[4]

How to strike back?

If You're Fighting a Superpower

1. *Cut prices!* Entice the client with the use of large, cost-effective tools, and the high productivity of your staff. Perhaps you and your client can jointly pay for expensive resources.

2. *Exploit the fact that at times the Superpowers are lumber-*

---

[4] Please note: Advertising won't help you win a specific contract. It's overrated as a direct sales tool for services companies.

ing giants, whose technology might not always be as up-to-date as smaller competitors'. The clone computers—many of which are faster than the IBM ones—are a good example. Tell how your computers will let the clients' managers whip through their spreadsheets faster than the IBM machines will. In general, Superpowers tend to be conservative in adopting new technology, new methods, new procedures—so that you can constantly ask yourself what newfangled computers or other tools could reduce not only prices but also delivery times, while improving quality of service. For instance, how about computer-aided design and computer-aided manufacturing (CAD/CAM)?

Also try making the ultimate argument against IBM, AT&T or equivalents: They might favor a solution using their equipment over one that would be better for the buyer. IBM might deny this to the hilt. Fine. Let it. But you're not in court, and both the buyer and rivals certainly have a right to consider a Superpower "guilty until proven innocent."

3. *Adopt the buyer, and stroke him.* That doesn't just mean a few poker games or fishing trips. Above all, it means helping him look good in front of his boss.

At no charge, lend him one of the pieces of equipment that you're going to propose.[5] Let him play with it. Ditto for software—in fact, even more so, since software takes time to learn, and since he can better appreciate the usefulness of the package if he can play with it. He can see for himself that your ware—hardware, software, or peopleware—will answer his problem. With such freebies, you'll significantly reduce the buyer's fear of failure, a far more effective strategy than losing to him in poker.

Also provide the buyer with a position paper outlining the arguments for your solution. Probably he won't agree with every word. But he can pick up the points he likes, and use them to influence his colleagues.

4. *Consider obtaining a subcontract from the Superpower.* You could augment its staff on proposals and projects in areas

---

[5] Obviously your loans should be within the law. You don't, for instance, want to be accused of bribing a federal procurement agent. That isn't exactly the best way to endear him to his boss.

where you have the experience and credentials. Seek out buyers and managers working for the Superpower—perhaps former colleagues—who would be willing to depend on you. Be careful, though. A Superpower could also be a Marauder and pirate people or plans, even take over your company against your will.

If You're Fighting for a Superpower

1. *Learn about new technologies, solutions, services, methods, and so on, that are not in your firm's repertoire.* See if they'll fit your company's image and goals.

Let's say you're in public communications. Then don't venture into customized computer systems for personnel management in the Department of Defense. Instead go into local and wide-area nets[6] for personnel management.

2. *If need be, set up subsidiaries that will be more willing to break into new markets.* In fact, since the selling of products differs in *most* ways from the selling of services, you might set up a new subsidiary anyway if you're just breaking into the services market.

Also think about establishing intrapreneurial companies where your individual managers can exercise their entrepreneurial, risk-taking instincts without clashing with your company's traditionalists.

Using either independent subsidiaries on a long leash or the intrapreneurial concept, you might even go after some of the niche markets traditionally relegated to the independents.

3. *At the same time, don't be afraid to shop for products and services from outside companies, including the small-fry, if your own company can't meet your needs. Develop lists of potential subcontractors that can help you break into the new technologies that clients love.* The example of the IBM PC—made at first mostly with parts from outside suppliers—comes to mind. The PC was a product. But the same concept would apply even more to the services industry.

Ideally, of course, you'll be using the outsiders not because your prices or quality lag but rather because of the other compa-

---

[6] A wide-area network lets a user talk to computers in distant cities.

ny's specialization. But what if your own products or services are losers? Should you use business from the strong parts of your company to shore up the weaker parts? In some cases you'll have no other choice. Especially in professional services, however, you should respect the concept of competition as the corporate reckoner. If divisions or their people are weak and you can't shape them up, then get rid of them. You'll be doing them a favor. It may be bad for morale in the short term, but people resent being welfare cases—which they are *if* they aren't pulling their weight and you can't retrain them or place them elsewhere. So sell off! Lay off! That's the best response if you regularly find that outsiders can give you better quality or prices than you could enjoy through use of in-house turkeys.

4. *Respond head on to the charges that your services branch will forget the client's interest in order to push equipment made in-house.* Never, never deviate from the policy of trying to offer the client the best solutions at a competitive price.[7] That's the best answer to the rivals who charge you with a built-in conflict of interest because you provide both consulting services and equipment.

Respond another way, too. Separate your services and manufacturing operations as much as you can, except when you sincerely think that clients will benefit through coordination between the two.

Notice the "except." If, by eliminating an outsider's profit, your own equipment will be cheaper to the client, you'll certainly want to use it. But make sure you're really serving his interests.

SUMMING UP

A true zero defect marketer will not succeed without allowing for surprises of the kind that killed Tom Finch's deal. Also, he must appreciate rivals' different styles and choose his tactics

---

[7] Of course there are times when the client might not want the best solution. IBM would be foolish to propose other equipment, for instance, if the client insists on computers from Big Blue.

with that in mind. The same would be true of buyers. As we'll learn in the next chapter, they, too, are adversaries; in fact, you must constantly consider the interplay between them and your rivals.

Do's

1. Outsmart your competitors by investigating their track records, strengths, and weaknesses. Find out what your rivals do better than you.

2. Raid your competitors' clients! But be sensible. It takes time to steal business from your rivals. Gird for long fights.

3. Protect yourself from raids—on your business or your staff. Stroke your clients as well as your employees.

4. When you deal with competitors, always work up a contract agreement to protect yourself against theft of proprietary information, staff and client raids, and other damage.

5. Appreciate the different styles of competitors and adjust your tactics accordingly.

Don'ts

1. Do not become obsessed with why you lost a contract. Instead try to figure out why the other guy won.

2. Never underestimate the power and influence of rivals in the client community.

3. Don't assume that your competitor's proposal is inferior to yours.

4. Avoid internal strife, which can distract you from your fights with rival firms and deprive you of the organizational backing you need for victory.

5. Don't think a competitor is a rival forever. In the future you might feel comfortable teaming up with the same firm—as long as you've protected yourself with written agreements and find that you'll indeed have mutual interests.

CHAPTER 6

BUYERS

Jane Wentworth failed to grasp the interplay between rivals and buyers—an oversight that cost her employers tens of millions of dollars when it lost a Defense contract.

Her buyer, a colonel, had been planning one of the Army's largest and most critical computer systems for national emergencies. And he wanted the system on-line as soon as possible to replace a haywire combination of computers and manual records. He didn't want to wait six years, the time his experts said it would take.

"We could do it in 18 months," said Wentworth. "We'll use automated design tools. Just as reliable as the regular way. And you'll have people using parts of it within weeks. What better way to get feedback on whether it works? We call it prototyping."

Wentworth's company won the contract to build the first part of the $400 million system. And sure enough, it performed as promised.

But Wentworth still lost out on further work because her rivals swayed Army away from her. Since prototyping was new, the colonel's engineers hadn't the foggiest idea of what was going on, and feared loss of control. They lacked the proper training. And the new approach didn't even comply with published standards—a point that Wentworth's competitors harped on again and again, behind her back. Had she spent more time educating the engineers rather than just wining and dining her favorite colonel, Wentworth might have won the job. Alas, she didn't know that some of the engineers were drinking buddies with her rivals. These competitors' livelihoods depended on Army's continued use of the older design techniques. Only through a fluke—the colonel's personal faith in her—had Wentworth won the small, preliminary contract in the first place.

How to avoid the sin that cost Wentworth the big contract? In this chapter, you'll learn:

- More about the basics of dealing with buyers, especially the relationship between them and competitors.
- How to gather intelligence on buyers.
- Ways to identify and respond to different styles of buyers, from the ignorant "Canon" to the well-informed, ethical "Fair Dealer."

BUYER BASICS

Your buyer is the source of the contract—the antelope breeder. He could be:

- A purchasing agent.
- A government contracting officer.
- The person actually using a computer system or other technology (the "end-user").
- His project manager.
- A consumer such as a small businessman.

Whatever the buyer is, he can't ethically guarantee that you'll catch an antelope. But it's like Snark, too. The buyer, the equivalent of the dealer, will pick the winner. And a fair-minded buyer can at least give you a good feel for the rules of the game, official and unofficial. You have everything to gain by becoming the buyer's friend.

Alas, one of the rules of the universe is that he starts out as your adversary—or at least not as your friend. You must convince him you're the best qualified and that you can reduce the risk of failure, both his and yours. Simply put, the buyers won't buy major and risky solutions from strangers. He prefers someone whose strengths and weaknesses he knows. And so an unfamiliar stranger with perhaps a promising solution might lose to a familiar bidder who, even in the buyer's mind, seems wrong for the job.

How to befriend the buyer, influence him, and protect yourself?

The issue goes far beyond simply bidding low and complying exactly with the buyer's specs. Ideally you'll *market* well before he even writes the specs! Within the limits of law and ethics, do everything you can to influence the contents of your buyer's statement of work—where he's outlining for his organization what he intends to include in the solicitation.

Never, never try to have the contents changed in a way that could harm your buyer. Either he or his boss may catch on. You can bet that your smarter rivals won't shy away from complaining about arbitrarily exclusive requirements. Don't risk making the buyer seem like a fool for awarding you the contract. What's more, a rigged solicitation could lead to bad publicity, either for you or your client; especially in the case of a government contract, bureaucrats talk.

At the same time—just as lobbyists write bills for congressmen and public relations firms write press releases for reporters—you may write an honest statement of work for your prospective client. He might not use it. But at least you'll have exercised your right to bring before him the issues you think are important.

In writing the work statement, you should consider your real strengths and weaknesses.

Say, your competition for a computer contract has a closer relationship to designers of IBM-based systems than you do. Then the work statement might emphasize performance or economy over full IBM compatibility.

Similarly, if only your firm has special tools for the proposed job, don't be bashful about specifying them in the work statement as a requirement. You aren't breaking the law. If you're up against the *more savvy* services firms—the few that do know how to compete—then they themselves will do anything they can to influence the actual work order. Why shouldn't you compete on a level playing field?

The same rules that cover the writing of work statements would cover research or research program plans. Ditto for equipment purchasing specs. Ideally, in the end, you'll be better able than the other lions to understand and respond to the document that really counts: the actual solicitation.

Mind you, even a skillful attempt to influence the solicitation isn't complete protection against competitors. That's because:

1. *A hidden competitor might still win out.* You're not the only firm in the world with the brains to influence or correctly interpret the buyer's specifications.

2. *If a buyer listens to your suggestions about the specifications, he may very well be similarly helpful with your competitors.* Maybe he's naive about high tech. But if so, you may not be the only expert with whom he'll consult. Don't be cocky. Don't think you really have the buyer in your pocket. Beware! Such a mindset can jeopardize your spirit of competitiveness.

3. *People other than those whom you've been lobbying may enter the picture.* You can, however, reduce the risks through good intelligence gathering and sound reactions to the information you uncover. Amass these facts as early as you can, ideally long before the actual solicitation. Find out who the other people are. What are their jobs? And how can they influence the awarding of the contract?

INTELLIGENCE GATHERING

Go to any good job counselor, and he'll tell you it isn't enough to answer a newspaper employment ad without finding out as much as possible about your prospective employer in advance.

Alas, many people in contracting act as if they're committing a federal offense by trying to get the facts on buyers and competitors. As a matter of fact, they would be in some instances. It is illegal to steal files. it is illegal to bribe people in the buyer's organization. Certainly you should consult your lawyer—or your conscience—if you harbor any doubt about the correctness of what you're doing.

You can do plenty, however, without either ethical or legal qualms.

Never, never confine your important bidding activities to a visit to the contracts office; that's the equivalent of a job applicant not going beyond the personnel office. A smart job seeker knows no advertisement can spell out what an employer will expect.

Similarly, a savvy bidder knows that a formal solicitation doesn't convey all. And just as the job applicant ideally will know of an opening even before it's advertised, so will a services bidder know of a forthcoming solicitation.

First, however, you should find out all you can about your buyer, his motives, and his attitude toward you. Some questions to answer:

- Question #1: Whom should your company really be marketing?
- Question #2: If your company has worked for the buyer before, have he and others in his organization been impressed?
- Question #3: What are the problems the buyer officially wants solved?
- Question #4: In general, what does the buyer *truly* want?
- Question #5: Do you or your competitors have a good, strong wire—an inside connection?

Question #1: Who's the Market?

Who is the true buyer, the true target of your marketing effort?

Simply put, you should try to identify anyone influencing the buying process—and accurately determine how much time you'll have to spend lobbying them. Granted, this book uses the "buyer" in the singular for the sake of convenience. But the word can refer to several or even dozens of people.

Never forget that buyers often will consult not only with their bosses but also with (1) peers and (2) people with whom your firm will directly work if you win the bid.

Question #2: Did the Client Like Your Past Work?

If your company has worked for the buyer before, have he and others in his organization been happy with the results?

Ideally the buyer simply will have told you so, but sometimes you may have to ask around.

Bureaucrats—public or private—aren't notorious for revealing their true feelings.

If you find that your buyer has been less than pleased, your proposal will have to address the resultant fears.

Say, you're an architectural firm and you recommended subcontractors who were too expensive. Then your new proposal should offer concrete evidence that this time you'll diligently shop around for the best price.

Question #3: What Problems Does the Buyer Officially Want Solved?

What are the problems the buyer *officially* wants solved?

Suppose the solicitation wants an office automation (OA) system to speed up correspondence production and financial planning. The buyer demands the most sophisticated of software packages. In addition, let's say, it asks for plenty of technical support.

That might be the official requirement. But you might have to ask yet another question. . . .

Question #4: The Buyer's True Wants

What does the buyer really want?

In requesting the sophisticated packages and technical support, he might actually want you to take over the correspondence and financial planning tasks. He might know that his own people lack time to learn the software packages well. All along, in fact, the buyer might have wanted to farm out the OA tasks. And yet he might have had to phrase his solicitation the other way to stay within company policy. It's happened, really. So ask about this possibility!

And also try to anticipate how your competitors perceive and will respond to what the buyer actually wants.

Question #5: Is There a Wire?

Has the principal buyer indicated informally that he's going to choose you as the winner?

That's all the more reason not to take anything for granted.

Your competitors may have picked up the same signals— and now may work all the harder to come up with better terms.

In gathering intelligence on the buyer, remember that you must always place it alongside the facts you gather on potential competitors. One set of facts will always influence the other! You can't consider the buyer and competitor independently. They're twin adversaries. Don't try to understand the buyer's style without also considering your possible rivals.

CANONS AND OTHER KINDS OF BUYERS

People and organizations don't neatly fit categories, but in reviewing your intelligence on your buyers, you'll find that some traits almost always stand out. So here I'll blithely proceed with my stereotyping of:

- "The Canon," a pure vanity buyer, dogmatic, ready for a huckster friend to cater to his ego.
- "The Brickbilly," indecisive and ignorant.
- "The Country Lawyer," seemingly a Brickbilly but actually a "twofer"—trying to squeeze from you twice as much as what he paid for.
- "The Dunner," a greedy one with some nasty penalty clauses.
- "The Auctioneer," a bargain hunter who doesn't want you to lower your price until his last round of shopping.
- "The Checker Player," who wants you to give him a break in return for his giving more business in the future ("everyone gets their turn").
- "Merlin," a power-hungry manipulator.
- "The Fair Dealer," the ideal kind for both you and the client organization.

Yes, most of the descriptions above are pejorative. So let me offer a caveat here. Don't think of the buyer as an adversary at all times. I myself have nothing but respect for the best buyers, the Fair Dealers. Also, never forget that many people in the services industries are themselves buyers as well as vendors. I shop for people. I shop for hardware, for software. And so I empathize with the buyers as well as the sellers, and I hope that they, too, will find this chapter useful. If the buyers can avoid the mistakes of the Canon and other sinners and gain reputations as Fair Dealers, they may attract more and better bidders.

The Canon

This kind of buyer is both vain and unsure of himself and tries to compensate by manipulating the rules to his own ends.

He may or may not be experienced, go to trade shows, or read professional journals. But one trait is certain—a steadfast belief in his own infallibility. He *knows* what he wants. And that counts more than anything—even more than the wishes of the people who'll ultimately use a computer system or other goods or services.

The Canon is deaf to other solutions. He might just throw you out of his office if you're foolish enough to suggest persistently that his specs are wrong.

Typically you've arrived there too late. A huckster—with a sales quota to meet—skillfully made up the Canon's mind for him. The insecure Canon has latched on to the salesman's solution.

Remember the example in the last chapter where a computer salesman sold the Defense Department an underpowered machine, which forced the programmers to work two shifts, six days a week? Quite possibly the buyer was a Canon.

There is one ray of light here. As mentioned earlier, the huckster's victim may think long and hard before giving the sales rep some repeat business.

Then again, the huckster might once more push the right buttons to make the egotistical Canon act out his dogmas anew.

Counterattacking the Canon
Some rules to follow:

1. *Even if you've arrived too late to make a sale immediately, don't give up.* With the future in mind, you should still market to the Canon. Give him a little free advice to get around the inadequacies of the machine or services that the huckster inflicted on him. And offer training or other services of your own.

2. *Try not to bid at all if you know that the Canon's continuing to insist on an impossible solution or set of specs.* He'll very possibly blame you when he fails. What's more, if he expected a bid from you, your lack of one just might help send him a message—that you're tired of his going off half-cocked.

3. *If you find that you must bid—perhaps because your boss says the possible contract is too big to ignore—then you'll have to offer the Canon's solution.* Yes, your own solution may be better. But it'll make you lose right away. Should you feel compelled to submit your solution anyhow, then you'd still better include the Canon's pet solution, with your own as an option. See Chapter 9 for suggestions for writing solicitations for the Canon and other kinds of buyers.

4. *Should you win by offering the buyer's solution, you'll have to shift responsibility for liability.* Bureaucrats and others love to beat up on contractors and blame them for program failures. So you have a tough job. In fact, as soon as you've won the contract, you might demonstrate your own solution vis-a-vis his—or at least do a study. Of course there's a danger. The defensive Canon will immediately feel you're attacking him. Whatever you do, cover your posterior—so you're safe if the inept Canon is transferred or fired. Ask the Canon to sign off on *everything*—on every problem report, every correction.

The Brickbilly

The Brickbilly might have grown up in the big city. But in his new job—or maybe even his old one—he's the stereotypical hillbilly. The Brickbilly is floundering. He doesn't understand the technology, the people or their games.

Imagine, then, how chaotic the Brickbilly's solicitation can be.

In writing the solicitation, he may have taken advice from everyone available at the right time—especially a huckster. This mishmash inspires a confused set of specs: a puzzle disguised as a solicitation.

Counterattacking the Brickbilly
How to respond?

1. *Consider not bidding at all.* If the Brickbilly doesn't know what he wants, he will probably blame you for failure and not adjust your profits on cost overruns to correct his deficiencies.

2. *If you do bid, make your response orderly and complete—but follow the solicitation's own organization.* Add charts and other explanatory material to create order out of his disorder—and include an assumptions and constraints section in your proposal. See page 203. If possible, try to educate the buyer before he sends out his solicitation. Or better yet, write the statement of work and specs for him.

3. *If you feel there's a possibility that the Brickbilly will unfairly hold you responsible for failures, then make him sign off on as much as possible when you're carrying out the work.* In that sense, the dangers of dealing with the Brickbilly might not be different from those of dealing with the Canon.

4. *Before submitting the bid, you should do enough research to allow for the fact that the apparent Brickbilly might instead be . . .*

The Country Lawyer

Remember the Watergate hearings? Sam Ervin, the senator from North Carolina, represented himself to the world as just a simple country lawyer—even though he held an Ivy League law degree. It was a delightful, entertaining form of reverse snobbery. And it may have thrown a few of his opponents off guard by making them underestimate Ervin's abilities.

The late Senator, of course, hardly held a patent on the country lawyer act—an art form practiced by many contracts officers who would make you think they're Brickbillies. But actually they know the rules of the game, all too well, and will extract more than you bargained for.

You can spot a Country Lawyer if his solicitation specifies candles and he gets you to deliver flashlights. Also, he will talk your fee down. He'll say he doesn't have enough money, or that he'll make it up to you at a later date. He's a manipulator.

Counterattacking the Country Lawyer
The rules:

1. *Fathom what he's really looking for, and write your proposal accordingly.* Is he an empire builder? Then offer him an optional approach—at your price—that provides a channel by which he can get what he really wants. Or else you should specify a limitation of what he's actually going to get. One way or another you're playing his game. And yet you don't lose.

2. *Remember that even though he might inform you of his vendor selection criteria, they may be vague.* If you can't puzzle out what he has in mind, you can still bid if you follow the advice above and offer what he wants—through an option available at your price. But you might be able to clarify selection criteria by talking to others who will evaluate your bid. Another approach is a detailed analysis of his demands. Does he want an economy solution or a deluxe one?

If you can talk to the other people or guess what the country lawyer wants, you might enjoy a 25 percent chance of winning. See page 74 in Chapter 4 for relevant tips.

The Dunner

The Dunner, like a bully, loves to throw in penalty clauses and otherwise try to squeeze out of you more than your contract requires. Usually he hates outside contractors. This empire builder would rather do the work inside with his own staff—people he can control.

Fall one day beyond deadline, make one more error than expected, and the Dunner will be knocking at your door—demanding either a penalty payment or extra services at a discount rate or even for free.

If the Dunner were shopping for automobiles, he'd insist you give him a new Rolls Royce for $500 with a 500,000-mile guarantee.

Unless you take suitable precautions beforehand, you may never win. He might even demand a refund if you finish ahead of schedule—because the job took less time than expected.

Or let's say lightning strikes your offices and they burn down. "An Act of God," you tell the Dunner. "The job will take a few weeks more."

"Then," he'll reply, "I think you should be penalized $2,000 a day for the delay."

"But the lightning rod wasn't working."

The Dunner flips to page 536 of the contract.

"Here it is," he says, and reads in a haughty, lawyerlike voice: "Contractor shall be responsible for inspection of lightning protection apparatus, whether owning or renting in said building."

Another trick of the Dunner is to ask for freebies such as free publication of documents, or no-cost travel.

Like the Country Lawyer, the Dunner looks more generous than he is. For example, the contract might provide you with free computer facilities. You bid based on that assumption. Too late—after you've signed on the dotted line—you learn that the offer won't apply except between midnight and 5 A.M. The miser doesn't want you to exercise your contractual rights. And very likely you won't. If you do, you'll pay your people a shift differential.

The ultimate offender, of course, might not be the government contracting officer or the private purchasing agent. It could be the people you're working with directly in the program. Either way, the result may be the same—an ostensibly lucrative contract turned into a red-inker.

Counterattacking the Dunner
Retaliate in advance:

1. *Think about increasing your prices to cover possible penalties or requests for additional goods or services.* If you can't raise your prices and still remain competitive—well, those are the breaks. Rejoice that a rival bidder instead wins the pleasure of being squeezed. But you still want to compete? OK, then specify your own conditions under which you'll do the job. For instance, you'd say that your costs were based on use of the computer between 9 A.M. and 5 P.M. Also, make sure you've asked what "free computer facilities" means.

2. *Prepare your proposal under guidance from lawyers, not just the usual techies and managers.* Stick in some crafty penalty clauses of your own—for instance, for failure to come up on schedule with the information you need to do the job. Or suppose the buyer must furnish facilities. Then, if possible, be very specific about when they must be available to help you meet the contract's work requirements.

Ideally, you'll be able to word your shyster clauses so that:

1. The buyer will be obligated to cover any contingencies.
2. He'll tear out his hair, trying to make sure that terms and conditions are definitive.
3. Next time, the buyer might not be so eager to clutter up solicitations with penalty clauses.

Don't feel embarrassed that you're tricking the buyer. He's already written the rules of this Snark game.

The Auctioneer

Auctioneers like to force prices down after the first round of offers come in. Then they make the rivals bid against each other individually and in secret.

Counterattacking the Auctioneer

To counterattack the Auctioneer, you try to find out how much he wants to shave off.

Don't answer the solicitation originally with your best and lowest price if the buyer indicates that he will be negotiating a best and final offer. The situation is very sensitive. Make sure that if conditions change, you will be able to submit an updated price before the final selection of a winner.

If you do bid your best and lowest price at first, and even if it is the overall low price, the Auctioneer might very well reject your bid because you haven't played his game. That is, you didn't let him cut your price.

The Checkers Player

The Checkers Player doles out business to a stable of companies, which might as well be pieces in a checkers set. He'll jump over you if it isn't "your turn" for a contract.

"But I'm the one with the lowest price and highest qualifications," you protest.

"Next time," the Checkers Player says. "It's Ed's turn."

Government and private industry could do without such fair-mindedness. For the Checkers Player is actually a talented manipulator, intent on squeezing the most out of the bidders. Here's why:

1. If your price is too high, he still might not recognize that it's "your turn."
2. He'll try to talk you into reducing your profits: "Look, do this for me and I'll owe you one, and you'll make out like a bandit on the next contract."

Counterattacking the Checkers Player
Respond this way:

1. *Try to refuse to let him play checkers with you.* It's too risky. Perhaps you should protest to his boss. Unless you have hiring and firing authority over the Checkers Player, you won't even know if he'll be around when it's your "turn." What's more, you may be transferred or move elsewhere yourself before your turn.

2. *Appeal to the Checkers Player's self-interest, while making it clear that you have enough self-esteem to value your own.* Say: "I'll welcome your business, but also your fair-mindedness. I want to look good with *my* boss—now, not just when it's my turn. It'll help you, too. Final costs are hard enough to calculate for you without worrying about favors and turns. Besides, in formally evaluating your performance as a negotiator, how many service firms will bother to ask, "Is it your turn yet?"

Merlin

Merlin, another manipulator, is a magician. He'll even raid colleagues' budgets to preserve his own projects.

On occasion, a famous Merlin will inspire fawning news stories praising his skills as "an entrepreneurial bureaucrat" who can get any program running.

Merlin is really an antihero. He finds loose equipment and loose change, and scrounges to get his projects off the ground. He does. And for succeeding he deserves some respect. Try to befriend him, even if, in the end, his magical powers may fade and his erstwhile admirers may condemn him for illegal use of funds.

Counterattacking Merlin
The rules here:

1. *Rejoice that Merlin will stand behind the programs he*

starts, but don't alienate yourself from others in his organization.
If you become identified too much with this power-hungry person you'll lose out if his abrasiveness causes his employer to boot him out.

2. *Realize that as ego-driven a person as Merlin may be also be a Canon, or worse.* Make him think you're taking his proposed solutions seriously—they in fact might be very good. A skilled bureaucratic infighter, if nothing else, isn't going to be dumb. Also, as with the Canon, protect yourself by having him sign off on problem reports and modifications.

The Fair Dealer

The Fair Dealer—and I'm not using that term in the political sense—is smart enough to value long-term relationships with contractors. He values his reputation as a good person to do business with.

Like the Checkers Player, the Fair Dealer may have a stable of regulars. Unlike the Player, however, he doesn't have favorites. He wants the best deal he can negotiate at the time.

Simply put, he keeps his word and is the best kind of buyer to deal with.

Counterattacking the Fair Dealer

Here's how you should respond:

1. *Remember that he may be a nice guy, but he still has a job to do—vigorously protecting the interest of corporate stockholders or the taxpayers.* You should attempt to drive just as hard a bargain with the Fair Dealer as you would with monsters such as the Dunner. The difference is in tactics. Be straightforward. Don't resort to trickery. Cherish this customer. Treat him right, and even if he can't give you more business himself, he'll spread word among those who can.

2. *Keep in mind that the Fair Dealer has enough self-esteem to resent interference from above, especially the political kind.* Granted, there are exceptions. If the Fair Dealer's own bosses are susceptible to pressure from the Hill, for instance, you may have no choice but to respond the same way. The Fair Dealer will understand.

3. *Don't be afraid to approach the Fair Dealer with new* *project ideas.* The Fair Dealer has enough confidence in himself not to automatically reject an idea just because it came from a company trying to win a contract. Taxpayers or stockholders in client companies can only come out ahead if Fair Dealers learn about new solutions to existing needs. The Fair Dealer may not agree with you. But he'll listen. Just make sure that you're marketing a new idea worthy of the Fair Dealer's time—that you are attempting to satisfy his needs, not yours.

Don't sell. *Market!*

If there's anything the Fair Dealer hates, it's a huckster.

SUMMING UP

Appreciate the interplay between rivals and buyers and decide how you can respond, once you've gotten the facts through good intelligence gathering. You might not know the identities of your rivals for a particular bid. But you can shore up your defenses with the interplay in mind. Jane Wentworth, for instance, as mentioned before, might have tried harder to "inoculate" the Army engineers against her rivals' bad-mouthing of her company's new techniques. Take nothing for granted. That's especially true when you're deciding whether you consider a contract winnable enough to pursue it in the first place. Read on "To bid or not to bid?"

Do's

  1. Listen—so you can understand your buyer's concerns, biases, wants, and fears. To listen is to know. And to know is to win.

  2. The more you show that you understand your buyer's concerns and needs, the more he'll trust you.

  3. You must stay in touch with the client continually whether you have an ongoing contract or not. That's how you'll eventually win repeat business. It's a basic too commonly ignored.

  4. Be honest. Cheat the buyer once and he'll never forget.

5. Always make assistance available—quickly—when the buyer cries for help. Another basic often ignored.

Don'ts

1. Don't expect buyers to be loyal or sympathetic to you alone. A competitor is always lurking behind the next bush.

2. Do not expect to be best friends with the buyer until you sign the contract. And even then he won't be sure.

3. Don't think that an especially attentive buyer is a sign that you've successfully wired the job or have formed a solid friendship. What he's really doing is comparing your responses with your competitors'.

4. Don't reject buyers' ideas or solutions outright—no matter how foolish. That can immediately close the door.

5. Don't commit the cardinal sin of being late to a meeting with a buyer.

CHAPTER 7

TO BID OR NOT TO BID?

Years ago I met a racetrack tout named Mike, a small, dark-haired man in the requisite flashy jacket.

"I have a system," he said with Bogart-like confidence. You hear the same words incessantly in the high tech services industry when people debate the odds of winning a contract.

Mike's system used the software between his ears to analyze his scrawls in thick, black notebooks. This database recorded a horse's win-loss record, its lineage, the history of the jockey, even what the weather was like on the days the animal had won or lost. "My system," enthused Mike, "makes sense of it all."

"Not bad," I said as we waited for the starter gates to swing open at the Belmont track on Long Island. "Now pick the winners of the next few."

"Sure, kid." Mike pulled out one of the trusty black notebooks. "It ain't foolproof, but it's close." I waited, briefly. With almost machinelike speed Mike gave his answer. "Well, Number Two, 'Puddin,' in the next race, is a definite winner."

I beelined for the betting window, gambled $2 on the dependable "Puddin," and returned to the stands. Inside my head I'd already spent my easy winnings.

The horses paraded out of the paddock for the next race. Suddenly No. 7, "No Quitter," reared up on its hind legs, pranced around, whinnied and jumped a few times. Mike watched, eyes widening.

"Hey," he said, "that's the winner, not 'Puddin.'"

We settled back to watch the race. Number Three, "Home Again," won by almost two lengths.

I offer the story to warn anyone expecting to find an easy system here. I promise I'll never market a $99.95 version of this chapter on a plastic disk; too many "No Quitters" and "Home Agains" are galloping around for me to allow for all possibilities.[1] Whenever someone utters the word *system* to me I think of the odds that really count—the chances of it working.

You must reconcile yourself to making some bad judgments. I have a good win rate, about half of the new clients I go after; yet it's obvious I'm failing a good part of the time. Plainly, zero defect marketing is a goal, not a realilty. Just the same, with care, you can do much better than the industry win rate of 33 percent (it's about the best most companies can do to win new business with new clients, without knocking off an incumbent).

And so in this chapter you'll learn:

1. *Lee's Law, a corollary of Murphy's Law.* Murphy's wisdom, memorized by all engineers, with the possible exception of those at Chernobyl[2], is: "If anything can go wrong, it will." Lee's Law is: "If anything can deprive you of a contract, it will."

2. *How to calculate chances of winning a contract.* The best bet is when your client wants to continue a current contract. Other situations may not be so encouraging. If you've heard that your prospective client is bad-mouthing you behind your back, for instance, your odds may be *lower* than 20 percent. In this chapter I'll tell how to quantify your chances through a numeric index.

3. *Assessing the chances.* Remember, the numbers you arrive at are guidelines—not absolutes.

4. *How to act on the probabilities and your intelligence and make the actual bid decision.* You may even decide to back away from profitable and *winnable* contracts—for reasons I'll explain.

5. *Two ways to control the bid process.* You should familiarize yourself with the effects of the two Cs—costing of the contract and the composition of your company—to help you pursue better opportunities.

---

[1] Perhaps some kind of an "expert systems" arrangement would do the job. That would cost tens of thousands of dollars, however.

[2] The ill-fated Soviet nuclear plant, the site of the world's worst nuclear accident.

LEE'S LAW

You know that Company A isn't a threat; it's too busy fending off a hostile takeover. Company B is too busy with other work to pursue this contract seriously. C is an ineffectual antelope catcher. No, the real one to watch is D. But even D's no match for you and your fellow lions, since (1) you really have a strong proposal and (2) you've been taking your client to lunch and losing ever so gracefully at poker. You're convinced the buyer loves you. In fact you believe that D will drop out of the running.

You bid.

And then? Oh, no. At the last minute the buyer leaves the company, and his replacement, a college classmate of Company C's marketing director, lets C steal your antelope.

Lee's Law has struck again.

I'm better than almost anyone in my field at calculating probabilities, but hardly infallible. Remember: Half the time I myself fail to win a contract. Sometimes I'm fighting bad odds but think the antelope is fat enough; but on some other occasions I'm a victim of my own Law.

Ahead are the five corollaries of: "If anything can deprive you of a contract, it will."

Lee's Corollary #1: Antelopes in Lion's Paws Sometimes Break Away

Your 90 percent prediction of success can plummet to zero at the last moment because a top executive vetoes the money budgeted for the contract. Lee's Law has struck again. It seems as if the man wants the money for his own project. Your woes could be worse, of course; what if the executive disbands the organization that employed your prospective buyer altogether? That might not be the best news for your plans to seek other bids from the same organization.

What's more, in government, Congress can slice the funds or assign responsibilities to other agencies.

Whoops! There go your chances of meeting your revenue goals.

Lee's Corollary #2: Antelopes in the Paws Sometimes Maul

You've been marketing this kindly, grandfatherly man for a year, and he asks you for an unsolicited proposal, which you spend $50,000 developing.

"Nice job," says the client when you present it. Clearly this antelope breeder is on your side. "Well," he says, "I'm going to put it through channels now for final approval. We should have some contract for you in four to six weeks." You *know* there's a 90 percent probability.

You call at the end of the month.

"It's still in channels," Grandpa says. "You know how corporate bureaucracy is."

Four more weeks roll buy. A friend of your calls from another services company and says: "Hey, I've got an RFP from the Grinch Corporation. Wanta to team up with us? It's really up your alley."

It is. When you visit your friend and look over the solicitation specs and statement of work, you find an exact copy of your unsolicited proposal of some months before. It even contains your budgetary estimate. So all the competitors know what you proposed, how you're going to do the job, how you scoped the work, and what you think the budgetary estimate will be. Whatever happened to your 90 percent odds? They're 30 percent now. Thanks, Grandpa.

He throws up his hands when you confront him. "Really didn't mean to harm you. But you know bureaucracy."

So much for the contract you were going to win with your $50,000 gamble. You feel as if this antelope is a giant carnivore.[3]

---

[3] In real life antelopes won't devour lions, obviously. However, other prey of lions—zebras—can maim lions with kicks in the jaw from powerful hind legs. The kings of the jungle then starve. A metaphor here?

Lee's Corollary #3: Antelopes in the Paws Sometimes Get Shot

A live antelope is more fun for a lion to hold than a dead one, especially if you consider the fact that a hunter who shot the antelope may have been aiming for you instead. Under such circumstances you probably won't want to hang around to chew on the antelope. And so you give it up. You might as well have not caught the antelope in the first place.

Constantly, bidders suffer rough equivalents of this lionesque trauma—seeing their antelope get shot.

For instance, you market a project for a year, aided by a teaming partner already enjoying several million dollars in business from the client agency. Your chances are 90 percent.

Then the shot rings out. Your teaming partner backs off and it's too late to find another by the bidding deadline—meaning that here again, your 90 percent is really a big, fat goose egg.

Lee's Corollary #4: A New Antelope Breeder Takes Over

You're one of two finalists in the negotiations when the phone rings. Partly because of the astute purchases that the buyer has made for his organization, his department is promoting him to an important job at the other end of the country. His boss is going there, too.

The new buyer isn't as careful with the tax dollars, and he tosses the negotiations out and begins all over again. The $50 million contract goes to one of your rivals.

Goodbye, antelope.

Lee's Corollary #5: A Crafty (or Clever) Lion Takes Your Antelope Away

A sneaky rival uses chicanery to learn your final bottom line figure, then successfully revises his bid price. Or perhaps there's nothing amiss at all. He might just have found a computer for the job that is twice as cheap and fast as the one that you have in mind. You were tempted to propose this alternative yourself. But you didn't hustle the way your rival did. You didn't think that your rival could surprise you by being so quick to take advantage of the new technology. Whatever happened, you're out one antelope.

Corollary #5 and the others all come from reality. With care, however, you might avoid most of the disasters described. For instance, to try to prevent Corollary #1 ("Antelopes in the Paws Sometimes Break Away") from kicking in, you might try to befriend people at the very top who potentially might make trouble later on. The same technique might also help protect you against Corollary #4 ("A New Antelope Breeder Takes Over").

PROBABILITIES OF WINNING

Your client mentions a problem. Wheels spin in your head. You start mulling over solutions; but simultaneously, you're pondering something equally important—your company's chances of winning the actual contract. At times the decision is tricky enough to need several meetings where you and colleagues ponder, "To bid or not to bid?"

Along the way your probabilities may vary wildly according to such factors as:

- Congress's mood toward funding a program.
- A corporate board of directors' feelings.
- Negative or positive publicity. If the newspapers are running headlines about your company's alleged overcharges for architectural and engineering services, this mightn't be the best of times to seek business from the General Services Administration.
- Changing technology. You might be basing your solutions on last year's technology and might not be allowing that the client is savvy enough to insist on the newest gadgetry.
- Internal factors, such as the resignation of a key employee whom you hoped would win your business.

When used in the absolute sense, probabilities aren't all that solid as predictors. Still, they count some. If your forecasts are overwhelmingly gloomy, you can back off before it's too late.

You should use probabilities early in the potential bidding process to help:

• Decide whether to invest in marketing and proposal development. You're not going to bet the company on a contract with a less than 20 percent probability rate.
• Predict potential company income within a period. The rule is: Expected revenues times probability equals weighed potential revenues. Suppose you have (a) a 30-percent chance or .3 probability of landing a $5 million contract and (b) a 50 percent chance or .5 probability of landing a $1 million one. Then, to calculate expected revenue, you'd multiply $5 million by .3 and add the product of $1 million times .5. Your potential revenue: $2 million. That's what you'll use to base recruiting goals and plan expenditures such as bonuses and facility expansion. You will *not* use the mere sum of $5 million and $1 million.

You can't, however, predict aggregrate revenues without first forecasting the chances of happiness coming your way on individual contracts. The answer? Create a numeric *index,* as a guideline to determine whether to pursue the job before you receive the solicitation (if you're sharp, you'll probably have known about the job well in advance of the actual request for proposals).

You'll generate the index by assessing a number of realistic conditions or factors that reveal your chances of victory. As the index scale is numeric—say, from .01 to .95—some companies can assume that the number also represents probabilities. Here's how this could work. A company will determine that an index of less than .25 represents little opportunity to win. This means that many of the relevant and important conditions that would contribute to a definite win are either incomplete or missing. You must look at the parts that drag down the average and see if you need more marketing to improve your chances.

How Winnable Are Different Kinds of Contracts?

Possible contracts fall into four categories, listed here, beginning with the most winnable and going on down:

1. *"Follow-on, noncompetitive contracts" are as close as any to sure things.* You're working for a client now and he wants to expand your current contract. This antelope is as good as eaten.

2. *"Recompeted jobs" are still promising.* Your client wants to check out the other companies but isn't necessarily unhappy with your work. And as the incumbent—familiar with your client's needs—you have an advantage over the competition.

3. *In "total open competition," your chances decline.* You may or may not know the client, the rules, or the strength of the competition.

4. *Going the "sole-source, unsolicited," routine might not be all it's cracked up to be.* Here you discuss a prospect with a client and he asks you for a proposal addressing the technical issues. But he hasn't promised that a contract opportunity will arise.

Follow-On, Noncompetitive

The Dole Company has been enjoying your services for years. Funds aren't any problem—Joe Dole could write out a check tomorrow. Your staff works so smoothly with his people that it's hard to tell them apart, and you yourself are engaged to marry Joe Dole's ugly daughter.[4]

Congratulations. Your chances of signing a contract are maybe 90 percent.

You can increase this slightly to about 95 percent if you're in the middle of a friendly negotiation—but don't count on the contract for sure until you actually sign. Who knows? Maybe Joe Dole's daughter has been having an affair on the side with a rival contractor.

In short, you never know when Lee's Law will strike. Consider what happened at Philistine Technology, which took for

---

[4] Or son.

granted that a client would renew a $20 million contract for a fourth time.

A Philistine engineer named Albert Barnes wanted to run for mayor. He'd been dabbling in politics for years, without harming his performance; but the Philistines still worried that his mayoral duties might interfere. Albert balked. After he won the election, Philistine Tech fired him.

Several days later, Albert's ex-manager visited a client to discuss a seemingly inevitable contract renewal.

"Well," said the buyer, "this may take time."

"I thought you were signing next week," said the Philistine.

The buyer coughed. "Ah, let's just say a little longer."

He was right. The contract didn't squeak through for more than six months, and meanwhile Albert's ex-boss had to close shop. The Philistines didn't grasp the real cause of the delay. Albert's brother and sister worked in purchasing and contracts administration and resented Philistine Tech's treatment of their little brother, the mayor.

Final Funding Approval Helps! At least Philistine's client had full financial approval of the contract. Even under the best of circumstances, the probability of a contract extension would be only 50 or 60 percent without a final approval of the money. And it might be as low as 10 percent without a budget and the client's technical approval. Time to scare up some business elsewhere!

If, however, the buyer has final financial and technical approval, you have a 70 to 80 percent chance of signing a follow-on contract. Indeed it might be 95 percent after all if Joe Dole's daughter keeps her engagement.

Recompeted Jobs—Follow-Up, Competitive

If you're doing follow-up work but must still compete for the contract, your optimal chances of more than 90 percent might drop to only around 50 percent.

This assumes that you're enjoying good client relations and technical approval and that the client likes your price.

Why not a higher percentage?

Well, that's often the nature of high tech services. Your client wants to make certain that he enjoys the *highest* tech—hence, his shopping around.

Technology, of course, needn't always be the reason. Within an area of competition, prospective bidders come and go, and market conditions change; so your buyer may simply want to see if he can drive a harder bargain this time around.

Open Competition

Careful here. If your proposal's lousy, your client might bad-mouth you to other buyers and say you're a loser even when writing proposals. Ditto if you're marketing ineptly.

You don't want such problems. It's as if antelope breeders are gossiping about a lion susceptible to kicks in the face from his prey.

So don't just write a technically solid proposal. Market! Try not to neglect any one of the antelope breeders—the people who, official or not, can influence the selection of bidders.

Count on 70 Percent Probability at Best if the Competition Is Open.

If competition is truly open, with at least several candidates competing, your probability will be no more than 70 percent at best. This assumes:

- The client has lined up funding.
- He really knows and wants you.
- You've "hard-wired" your client to your company—by writing the statement of work and the specs.
- You know all the competition, how they cost out their work, and how they price their jobs.
- You don't fear your rivals—perhaps because several of your top managers once worked for all of them.
- You have the best proposal and the best price.

Even here, because of Lee's Law, the true probability might be only 50 percent.

If Conditions Aren't So Optimal.

Your chances will sink to 30 or 40 percent with open competition and several competitors under the following conditions:

- You almost know who your rivals are.
- The client *may* know you.

- But you haven't worked for him before.
- Funds probably aren't committed yet.

Almost the Pits: When the Client Doesn't Know You at All.   You'll have less than a 20 or 25 percent chance of winning if both of the following are true:

- The client doesn't know you at all.
- You're qualified to do the work.

Unless you can get the client to know you in a hurry, you might as well drop out of the running.

When It's Time to Forget This Antelope.   Your chances are less than 10 percent—maybe even .01 percent—under these conditions of open competition:

- You don't know the client.
- He doesn't know you.
- You don't really know your rivals.
- But you're almost sure the contract is wired to someone else.

The Unsolicited Proposal

An unsolicited proposal is what it sounds like. It's your written or spoken response to your client's oral request for a proposal. Perhaps he's asking for a solution to a problem.

Normally an unsolicited proposal won't lead to, say, a $4 billion program—but it can lead to many small contracts, especially the research and development (R&D) variety. Or it can lead to follow-on work, either as sole-source or competitive contracts. You'll want to find out if the buyer has discretion to issue sole-source contracts.

"Sole source" means that the prospective client won't put the work up for bids, either because you're obviously the most qualified, or it wouldn't be economical to search elsewhere, perhaps because the size of the contract is so small.

Very likely such contracts won't have to go through the organization's usual budgeting and planning process. The discretionary fund category already covers them.

Unsolicited proposals are never the main source of revenue and contracts for larger services firms. But many smaller ones depend on them heavily. Typically the dialogue between you and the client goes like this.

"All the librarians are yelling at us because the computer's always swamped on Saturdays," says the prospective buyer, a data processing (DP) expert with a county government. "The lines are a mile long at the circulation desks." There: He's told you the problem. You nod. Your wife waited 30 minutes the other day to check out the latest Judith Krantz opus.

"I know how you could tweak the software," you say. "A new technique. You won't have to tinker that much with the system."

The DP man's eyes brighten. "We have some discretionary funds. Tell me, how would you do the tweaking?"

Your experts submit a written proposal, and the DP man likes it. Are you home free? No. Alas, one or more of the following may take away your antelope:

- The client might turn your private, unsolicited proposal into a public document for an open competition.
- The DP man may have a harder time than he suspected, prying discretionary funds loose, what with the county government's recent economy campaign.
- He might not have conferred with all his colleagues about the proposal, and they find that the end users, the librarians, have objections.
- Your client decides belatedly that you haven't sold him 100 percent on your solution.

Your probabilities might now be a mere 20 percent or even less. Determine the probabilities by way of this chapter's subsection called "If Conditions Are Not Optimal" (see page 137).

ASSESSING THE PROBABILITIES

Remember—probabilities are just guidelines, not absolutes. Don't let this warning insult you. I've made similar mistakes myself.

The way to overcome the risks is to think of probabilities just as corrective monitoring tools. That is, they show the extent to which you must do *something* to win that contract.

Say your computed probability is a mere 50 percent. Simply by better acquainting the client with yourself and your work, you might boost your chances to 70 percent—just before the buyer issues the solicitation.

Again, however, watch out for Lee's Law. The down side is that your known probabilities can plummet as rapidly as they can rise. Take what happened to the Hipp Company and its potential subcontractor.

> Hipp, a New York engineering and professional services firm, wanted to provide software and various computer-related services to a large government agency. Philistine Technology was the existing contractor.
>
> Privately the bureaucrats at the agency were cheering on Hipp. Philistine was arrogant. Its staff analysts jerked around employees of the agency. The bossy Philistines might as well have been the clients rather than the offerers of the computer services. But now the contract was up for a competitive rebid. Ah, revenue! Hipp or another Philistine rival seemed destined for the $30 million job.
>
> What's more, Hipp and its potential subcontractor, Opus Systems, had both hired away staff from Philistine's government contract staff. Hence they knew:
>
> - The strengths and weaknesses of the people drafting and pushing the Philistine renewal.
> - Philistine's pricing strategy.
> - The agency's problems—including those related to Philistine itself.
>
> Encouraged by the Philistine alumni, Hipp assigned the contract a 60 percent probability. It and Opus mounted a strong marketing campaign—including briefings for the agency's people whom the arrogant Philistines had slighted. The two companies invested $300,000 in the proposal. It was a Cecil B. DeMille production, tying up dozens of high-level people at both Hipp and Opus.
>
> Thirty days after Hipp plopped the bid down on the buyer's desk, the bad news came.

Philistine had triumphed.

Now, the Philistines in the past had made their share of blunders, but in this case they'd been both decisive and effective. In effect they told the agency bureaucrats, "We goofed. You be the bosses. We'll listen to you."

So in the end the agency's working stiffs went along with the decision of the brass to renew the Philistine contract. "They know you," said the agency's senior managers. "They know your computers. The Hipp people can't even find their way to the tape library."

Alas, while communicating closely with the troops, Hipp hadn't paid enough attention to the brass—and hadn't anticipated the Philistine's skillful response. The supposed 60 percent was zero.

MAKING THE ACTUAL BID DECISION

So far your probabilities have justified your marketing efforts and other investments in the potential job. Now you might be about to catch the antelope. A big, thick solicitation lies on your desk, perhaps; and you pick up the papers, laze back in your chair, and ponder whether to continue the chase.

Up to this time you've spent thousands of dollars gathering intelligence on your client and your rivals. You've taken the buyer to lunch. And dinner. And more. He's an early riser, one of those sadistic rise-and-shiners who wants the rest of the world to join him at the breakfast table at 6:30 A.M. And you've wanted that contract badly enough to go along. What's more you've pleaded with past clients to give you positive references, and you've persuaded your mother-in-law to call up her boyfriend whose destiny in life was to end up on the corporate committee that will evaluate the bids. Now you're hoping that your dreams will come true, that your optimism is justified, that you'll be more accurate than Mike the Tout.

But what to do? Do you put everything in numbers? Or will intuition be the key?

Well, be wary of those statistics. Your marketing department, together with optimists on your technical staff, may have helped you crank out those enticing numbers. But now the plan might be under the harsh scrutiny of:

- Your president.
- Cognizant line vice presidents.
- Quality control departments.
- The contracts department.
- The legal staff.
- The proposal publications and editing department.
- Marketing executives with no previous connection to the matter.
- Senior technical people.

Careful! Legal, for instance, may decide that the buyer's request for proposals contains too many clauses with stiff penalties for alleged failure to perform. "Is a $2 million dollar contract worth chasing," they may ask, "if the buyer can casually sock you with a hefty nonperformance penalty? Perhaps running up past $750,000?"

So see if you can't outguess your company ahead of time—by skillfully weighing both the positive and negative possibilities and conditions *and* by using intuition.

Bear in mind the weaknesses of this numerical approach. We all know how engineers can become obsessed with figures to exclusion of all else. Thank God. If they didn't, then bridges would tumble; skyscrapers, collapse; planes, crash. But even a Phi Beta Kappa engineer from MIT can't infallibly quantify commerce and human relationships, especially those that count in bidding wars. You never know what will matter most in the end.

Enough perspective? Fine. Your scoring will go from 0 to 10, no negative scores.

In following the lists of "positives" and "negatives" below, carefully review the condition implied by the questions for each *bid factor* (the numbered items).

Determine the extent or degree to which each condition is valid or certain. High degrees of certainty should be given scores from +8 through +10. If certainty is not very high, or not sure, then assign scores from +3 through +7. And if the condition is not true or not promising or certain, then assign scores from 0 through +2.

Now add up the scores for each bid factor and divide by the

number of conditions (questions answered) to get the score for the bid factor. Positive and negative conditions listed will both relate to the same bid factors.

Next add all the bid factor scores and then divide the sum by the total number of factors considered times 10 to obtain the final index. (See Appendix A for the Bid/No-Bid checklist to use.)

Here's an example of the formula in action. Let's say you're working on the "Bid Factor" of "Business Opportunity." And you're trying to determine if the contract is a prime target with minimum risk. You have been tracking it for a year, but there is some risk, so you assign a score of +9. Also you know that the contract has funding but that the client needs approval for issuing funds, so you assign a score of +8. However, from the negative side, you were surprised by the RFP or solicitation, which means that you missed something while tracking the program, so you assign a score of only +2. Thus: $(9)+(8)+(2)$ divided by 3 = 6.33. Sorry. The negative condition downgraded all the positive, high-score ones.

Positive Conditions

Here are questions to mull over when you're considering the positives:

1. *How attractive is this business opportunity?* Is the contract a prime target with minimum risk? How about funds? Can the client casually write out a check?

2. *What kind of contract is it?* One with a high potential follow-on? What are the chances of a good profit? Is the contract a cost-reimbursable one with no risks?

3. *How well do you know the customer, and vice versa?* Does the customer want to give the contract to you sole source? Are you in a strong competitive position? Did the customer request you to do the job?

4. *If you're thinking about subcontracting some work out, would this benefit both you and your prospective client?* Will the subcontracting improve your financial and management standing? And is the risk low that the client in the future would forsake your company and deal directly with the subcontractor?

5. *How well have you prepared your proposal?* Do you have an experienced proposal manager and a highly qualified technical staff—people whose work will respond exactly to the solicitation itself? Are they capable of remembering that the proposal very likely will be part of the actual contract? Can your company meet the technical, managerial and financial requirements?

6. *Is your company technically superior overall, compared to your known rivals?* If so, and if you make the buyer aware of this, you'll go a long way toward nullifying your client's fear of failure.

7. *What about your general capabilities and qualifications?* Do you have strong in-house contract and supplier experience for delivering equipment?

8. *Can you take on the new contract without harming existing or planned contract work?* What staff will be available? Will the people come from contracts being completed? Or from staff now on overhead?

9. *What about your ability to commit the right people to the project quickly?* Can you easily commit a highly qualified manager and technical people to the project?

10. *What facilities are available?* Is office space available—ideally near the client's site?

11. *What about your cost estimate?* From your intelligence—your sense of what the client wants and what your rivals may bid—have you been able to develop a winning price?

Negative Conditions

Now consider a second list, which will lay out the negatives; the general categories of questions will be the same, but the possible answers will differ.

1. *How attractive is this business opportunity?* Did the RFP startle you? If you weren't expecting it, you might not be as much on top of that market as you thought you were. And so you might have a harder time coming up with a proposal to meet the buyer's exact needs. Perhaps, too, the project's penalty clauses and other potential liabilities are too great. And funds aren't available. Be objective. Be prepared to assign a low score—for instance, 0, 1, or 2—if you've been napping and your victory chances don't look great.

2. *What kind of contract is it?* Is it perhaps beyond the scope of your company? Will it require a big investment? Are the specs too vague to cost it out for a fixed-price contract?

3. *How well do you know the customer, and vice versa?* Does the client have a current contractor with qualifications like your own? Is the incumbent wired in? If you subcontracted some work out, would both you and your prospective client suffer? Would your client be more ready in the future to think of alternatives to your direct services?

4. *If you're thinking about subcontracting some work out, would this harm both you and your prospective client?* Perhaps that's evidence of risk—suggesting you aren't qualified for the work. What's more, your client might decide in the future to deal directly with the subcontractor.

5. *How well have you prepared your proposal?* Are you lacking a good proposal manager and staff? Is your company missing the expertise to make the proposal truly responsive to the buyer's needs? What about other internal problems that could harm proposal preparation?

6. *Is your company technically superior overall, compared to your known rivals?* Perhaps the job requires certain engineering skills that your firm lacks.

7. *What about your general capabilities and qualifications?* Maybe you haven't any experience to do the work. That in itself normally would be enough to justify ending the chase.

8. *Can you take on the new contract without harming existing or planned contract work?* Will current contracts be in default if you borrow or steal their people for the new contract?

9. *What about your ability to commit the right people to the project quickly?* Your best manager and technical people for the job may be tied up for the next several years.

10. *What facilities are available?* Perhaps you can't find affordable office space within convenient distance of the client and the size of the contract doesn't justify letting a landlord gouge you.

11. *What about your cost estimate?* Does your intelligence indicate that the client might balk at your price, or that your rivals can easily undercut you?

More on Interpreting the Score

Let's say you've calculated the final results of the formula in all of the 11 bid factor areas, and your weighted average is, say, .5. This means that you have a *chance* of winning. Still, you need to try to change many factors and conditions to improve your score.

If your score is, say, .2 and the issuance of the RFP is imminent, then you should probably forget the pursuit. I said *probably*. There may be one or two very strong factors, such as incumbency and client relationship, that may really be worth more than your top score of 10. They just might convince you to bid. And you might even win.

Anyway, with a score of .2 and maybe six months before the official solicitation is issued, you have an opportunity and the time to increase the value and possibility of changing a negative factor.

TWO WAYS TO IMPROVE THE ODDS IF YOUR HEART IS REALLY SET ON WINNING MORE CONTRACTS

You can more confidently vie for business if you know the effects of the two Cs—costing of the contract and the composition of your company.

Each C can influence the other.

Costing

When you cost, you're setting the competitive price of your services for specific work at a specific time.

Keep in mind the word *specific*. You aren't selling soap or Oldsmobiles; you're not dealing with absolute market prices for a whole industry. Even prices for consumer products will vary, of course, but not so much by the customer.

Here are variables that can influence each other in the services industry:

1. *Supply and demand.* The curves won't be as neat as in the consumer industries.

2. *Price elasticity.* Lower prices may generate more business—but not to the extent that they can in the selling of a consumer product.
3. *Your own costs.* I'll explain the baseline costing technique.
4. *Your rivals' costs.* How can you intelligently guess them?

Supply and Demand and Price Elasticity

Some whizzes in labs invent a new technology. People of a more practical bent adapt the principles—to smarten up telephone switchboards, electronically run buildings, or automate medical gear. Suddenly the whole world wants to enjoy the discovery. It may save money, even lives, and no one wants to fall behind—whether it's a PBX user or a hospital keen on using the latest diagnostic techniques.

The new hardware creates a demand for related services from those familiar with the technology. Wages rise. So do companies' costs, especially as the experts job hop. Meanwhile, more services companies are jumping in, hoping to make millions off the techno-miracle.

In the consumer marketplace, however, other truths prevail. You drop prices to increase sales volume and cash flow; that's the whole idea of price elasticity.

But service companies' prices are another animal. They don't rise or decline so significantly, even when many bidders vie for business. Granted, if you do lower your price, you might win a services contract. Then again you might lose because the buyer questioned whether you appreciated the cost of meeting his needs.

Furthermore, in high tech services, the elasticity idea can't help you because a low price won't increase your market share. You've only influenced the outcome of one contract.

The best explanation for less elasticity, however, is something else: Normally bidders pay about the same for a staff, and they usually enjoy the same discounts for the required equipment.

Your Own Costs

Just the same, even without the cost elasticity present in consumer products, you can underprice your rivals.

Do so by way of baseline costing. It can help you better understand what margins you'll need to make the job worthwhile.

For people-related costs, calculate:

• Direct salaries.
• Overhead (including office space), general and administrative costs, and fringe benefits.
• Profit.

Next determine the "semifinal" price, based on:

• The job's potential for future expansion.
• The profit potential, based on your company's goals and needs.
• The company's image (are you known as a Cadillac within your field or simply for low prices?).
• Potential risks, especially those possible from liability clauses, as well as those that might damage your company's image if this turned out to be a problem contract.

Next assess the price in terms of:

1. *Your best guess about what the competitors' prices might be.* Remember that even if the rivals' costs are the same, the bid could still be higher or lower.
2. *Whether your price is within the buyer's own estimates or budget.* Avoid too *low* a price. Otherwise your buyer may worry that you don't know what you're doing, or that you'll rack up an overrun, threatening his budget. If you still want to be a price-buster, then tell the buyer why you're so inexpensive. Perhaps you're a smart shopper for equipment or the silicon chips in your proposed system will be newer and smarter.

A high price, of course, is even more risky; the buyer will omit you from serious consideration.

And if *all* bids are too high, he might go out for rebids or even cancel the contract or start heavy negotiations with a favored contractor. But watch out if you do learn your client's budget. People love to bid just below budget, but competitors without such inside goodies may bid even lower.

A Global Perspective on Pricing

Summed up, pricing involves the value of the exchange to the buyer.

In the end the main questions will be whether your price is consistent with:

Your offer.
Your rival's pricing.
Your buyer's budget.
The composition (and size) of your firm.

That is the topic discussed below—the second C in "cost and composition."

Composition of Your Company

GM and IBM normally chase different kinds of contracts than do the smaller companies—and it's no accident.

The biggies have contracts departments, legal departments, and all the other auxiliary departments that the jargonists love to describe as "infrastructure." The tech people who work for the client have many support staffers behind them. This of course will boost costs. However, on large contracts, clients' terms and conditions require that a bidder retain such an infrastructure.

The composition and functions[5] of a firm, in other words, will influence its size.

A large firm's composition can work either for or against it. On one hand, clients, especially large ones, may appreciate the additional care and professionalism that the best support people can offer—for instance, better edited manuals. Still, an infrastructure is expensive. A vast army of technical editors will remain on the IBM payroll regardless of the number of projects requiring their services. And a $100,000 a year Harvard-educated lawyer won't take a vacation just because a routine contract doesn't justify his time. In fact, those extra lawyers and

---

[5] Functions would include the kinds or categories of work that a company does for itself or clients.

other paperwork generators can actually complicate the client's life—even if the buyer is a giant federal bureaucracy.

Is it any surprise then, why some large companies won't chase after smaller contracts? The biggest companies, in fact, may even use $50 million as a cutoff.

And so the small-fry have plenty of business to themselves—as long as they're aware of their own limitations. Pratt Systems, however, wasn't.

> Pratt—20 people strong, mostly engineers—marketed a small engineering design contract to modify some government buildings. The feds' budget here was $110,000. "We haven't worked for this agency before," Steve Pratt told his people. "Let's go in at $60,000. We'll get an entre, get more business."
>
> Everyone congratulated Pratt on his boldness and worked overtime on a superior proposal.
>
> The contract, however, went to a competitor three times the company's size, and the deeply frustrated Pratt heard the hard truth from the buyer and his contract evaluation committee. "Fact is," said the government man, "we wanted to use you. But that $60,000 was a problem. We don't think you understood the job. And frankly, Steve, we think we saved you from yourself."
>
> "I can do without such rescues," said Pratt.
>
> "You see, this was a fixed price contract," said the government man. "If you'd botched up your schedules and deliverables—why you'd be risking a trip to bankruptcy court."
>
> "OK, so maybe I gave you a low price hoping you'd appreciate our work."
>
> "We already know you're good," said the buyer. "Let's just say we think buy-ins should be left to people who are big enough to play the game."
>
> The government buyer was right. The contract was worth some $7.5 million altogether, more than three times the yearly sales of Pratt Systems.

For a small services firm, especially, contract value should be not much higher than two and one half times a year's income. Marketing cost of a contract should not exceed .02 to .05 percent of the sales value[6] of the contract. However, if you are breaking

---

[6] Sales value would include all the income, revenue, or cash that a contract would bring in.

into a new technical area with a new client base and tough competition, marketing costs could go as high as .15 to .20 percent of sales value of a contract. (Note: The cost of a proposal should be about .02 percent of the booking value of a contract. A booking value is the figure you negotiate on the contract. It might not include the possible value over the contract's life.)

Financial complications aren't the only reason not to overreach. For a $7.5 million contract, a company would need to hire 45 professionals in 90 days—no small trick in a tight labor market. What's more:

- A small company mightn't be up to the task of managing the extra people and overseeing their work.
- Questions might arise about the contractor's stability— especially if key people left or suffered heart attacks.
- A small company might not be up to the technical challenge.

In the role of buyer, Philistine Technology suffered the frustrations of dealing with companies too tiny to do the job.

Philistine was to design a records management system for Brobdingnagian Insurance. And so the high tech firm shopped around for a subcontractor that could build a system using both computer-based digital records and the microfiche kind. That was asking plenty. At the time, the mid-1970s, normal computer systems couldn't directly find the goodies buried in microfiche. Philistine budgeted $2 million for the subcontractor capable of such a miracle.

The request for proposals drew several bids—two from large manufacturers and one from a small shop in the Midwest, which, with the lowest price, won. Just three professionals worked at Lilliputian Systems.

They took seven months longer than scheduled to modify the computer, and even then it operated at just 80 percent reliability—far below the 98 percent required—due to hardware problems. Unknown to Philistine Technology, one of the Lilliputians was quarreling with his two partners and he quit. And he happened to be Lilliputian's only hardware expert. The result? Philistine had to cancel the contract—losing time and money, and paying hundreds of thousands of dollars in penalties to Brobdingnagian Insurance.

If you're a Lilliputian, how do you avoid overreaching? One word: Imagination.

Imagine all the details that a larger contract will require. For instance, if your contract requires you to supply equipment, will your people have the experience to win discounts from suppliers? Do they have the financial know-how—or the corporate collateral—to borrow at the most favorable rates? Can they grasp the subtleties of, say, OEM (Original Equipment Manufacturer) agreements or value-added reseller agreements? If not, you might be open to law suits. How much better to be large enough to have finance men and attorneys working on your side from the very start so you can avoid difficulties later on! If those backup resources aren't available—either in-house or outside—you'd better forget about a bid and move on to another contract.

There are solutions of sorts, however. Search for professionals who'll join your company or work on retainer if you do win a contract larger than normal. Also think about teaming up with a bigger company. Subcontract work or be a subcontractor yourself. Yes, you'll get less money than if you bid alone, but you won't be jeopardizing your reputation for sound judgment. Don't chase too big an antelope single-handedly, lest *it* maim *you*.

SUMMING UP

Try to stay ahead of your adversaries in all respects when you ask yourself, "To bid or not to bid?" Consider the possible pros and cons, such as:

- Profit.
- Foot-in-the-door opportunities.
- Expansion.
- Compromise of other company commitments.
- Other risks, especially penalty clauses.

Also, you should ask:

- Do I really need the contract at this time?
- How good are my chances of winning?
- Realistically, what are my strengths and weaknesses?
- What is the probable competition?

- How do I compare on those strengths and weaknesses?
- What should I do to overcome my weaknesses?

Along the way, try to avoid being misled by myths. In the next chapter you'll read about "Nine Marketing Myths and How to Overcome Them."

Do's

1. Build up an adequate backlog of promising leads in case several seemingly good prospects don't come through.

2. Consider bidding on a job where you don't have a big chance to win, but where you could use your proposal as a public relations tool to lure business your way in the future.

3. Continue to use the bid/no-bid check list (see Appendix A) not only to determine your chance of winning but also to identify the steps you must take to assure that win.

4. Keep an intelligence-gathering team in the field to monitor changing conditions, rules, different players and any other factors that can influence your chances.

5. Periodically hold bid/no-bid meetings for all high-probability prospects even before the solicitation comes out. Then inertia won't push you along in pursuit of an unwinnable job.

Don'ts

1. Don't bluff in a Snark game. Too low or high a bid won't fool anyone.

2. Avoid bidding on a job where the incumbent has a Royal Flush—an ultimate advantage—showing.

3. Do not throw proposals over the transom just for the sake of writing proposals. Don't respond to every one coming down the pike.

4. Don't expect to get a second chance when you lose a bid. When you're out, you're out. If Agency A doesn't choose you to design and build a computer system in 1988, don't think you'll triumph in the competition to build a replacement system in 1998. Chances are that the original winner will have entrenched itself.

5. Don't waste valuable resources chasing too small or large a bid for your company size.

CHAPTER 8

NINE MARKETING MYTHS AND HOW TO OVERCOME THEM

Don't fall for nine popular myths that can lead you astray when you're seeking new business or deciding whether to bid.

Not every firm swallows every myth, of course. Still, too many people in high tech services go for old chestnuts, such as the belief that jobs announced in *Commerce Business Daily* are all wired. Ditto for those in the "bids and proposals" sections of local newspapers.

Myth #1: "Jobs Announced Are Wired"

If you defy this folklore—Myth #1—you might bid yourself and win. That's because prospective competitors often think an ad means it's just plain too late to bid. Mightn't a rival have marketed ahead of them? Obviously it happens. Large companies, for instance, are always chasing after major programs about which public budget documents have alerted them two or more years beforehand. But government buyers say that often contractors don't submit any proposals on small- and medium-sized projects—even when ads run. The buyers investigated this apparent phenomenon. Why did prospective bidders—who had asked for the solicitations before they saw the advertisements—back off? "Well," came the reply, "when it's announced in *Commerce Business Daily,* it's too late."

Thank God that many experienced marketers still believe this foolishness. Now, be a smart lion and chase after all those unclaimed antelopes!

Myth #2: "The Project Is Wired to a Competitor"

What a cop-out! Yes, thousands of wires run through Washington—and also through the corporate world—but *usually* you still have a chance. A superior proposal with superior marketing can win over a favored competitor. It isn't honesty that can cause a buyer to unwire a contract; it's fear—that someone will criticize him for not choosing the best solution.[1] Sometimes, moreover, Lee's Law can work in your favor.

Now, don't depend on it. But don't back off if you indeed can write the perfect proposal for an obviously superior solution.

Myth #3: "Hey, It's Wired to Us"

See #2. There's no guarantee, even if you did help write the RFP. In fact, you might have to work twice as hard—to help the buyer justify his selection of a favorite.

Myth #4: "The Client's Unhappy with the Incumbent"

Well, then, if the client's so unhappy, why hasn't he just cancelled the incumbent's contract? There are degrees of discontent. What's more, the client's staff may be at odds with each other over whether to renew, and the proincumbent faction may win.

Also, the incumbents often have time to recover from negative propaganda or bad management practices.

Myth #5: "The Job Just Can't Be Done"

Maybe it's a big solicitation or it's hard to decide what your costs will be or you really can't meet the specs.

Within the limits of realism, however, try to rise to the

---

[1] The disappointed bidder might even lodge a complaint with an audit office or organization—for instance, the General Accounting Office.

occasion; this is your chance to shine. Is the job too large? Then search out a bigger firm to team up with.

Is the job too hard to cost? Intensify your research.

Can't meet the specs? Try to restudy the potential client's needs and see if maybe he'd allow a different solution from the one expected. Of course you'll normally want to respond as closely as possible to the specs in the solicitation.

Myth #6: "We Don't Have the Staff to Work on the Project"

You're guilty of misdirected ethics if you refuse to take advantage of the resumes of people already on staff for proposed work. After all, current staffers are the major resources you can use for getting new work, although, of course, when you win the job, you can always hire new people. This is how companies grow. Diogenes, Inc., had flat income for seven years because the vice president wouldn't let the company use the resumes of staff already working on current contracts.

Then see if you can't pay your people overtime to whip out the proposal.

Also try to find outside people who will commit to work on the contract, and whose resumes you can use in the proposal. This is neither unethical nor illegal. Just make sure those extras are reliable and have a track record for keeping their word.

Myth #7: "We'll Propose (or Offer or Assume) More Than the Client Requests in the RFP"

You think, "Why not increase my chances of winning? I'll add new or alternative goodies."

Sorry—that's normally verboten. Your client is ready to review proposals within his specifications, not those dramtically beyond specs. By the rules, he'll evaluate just A and B. Not C, D, E, F, and the other unrequested goodies. Careful! He might even think: "You dummy! How can you do the work if you're too stupid to follow the instructions in the solicitation?"

Furthermore, he may mutter to himself: "The son of a bitch thinks he knows my needs better than I do. Well, I'll show him!

That RFP says what needs saying! I'm not about to dole out $5 million to a company that thinks I botched the job."

So from a business point of view—and a legal one, too—you should answer the RFP as closely as you can. Worry about the improved approach only after you win the job. *Then* you can put in a word for C, D, E, and F.

Myth #8: "The Client's Proposed Approach or Solution Is Wrong"

Bottle up your protest! Do not use the proposal as a platform to attack the client's approach or solution. Respond as best you can to what the client wants. But if you do want to protect yourself, use one of two ways:

• Way #1: Offer an alternative approach with discussion of benefits in addition to responding to the original specs.
• Way #2: Offer an "assumptions and constraints" section to set up protective mechanisms. See Chapter 10 for how an A&C can reduce such unpleasant possibilities as an unworkable solution.

Myth #9: "If We Bid Less Than Our Cost with the Lowest Price, We Can Always Win"

Your client, alas, is wiser than you think.

Bid too low, and, as I've said before, he'll think: "The idiot really doesn't understand the scope of the effort." Or, "He's going to lose money on this project and we'll have a nightmare renegotiating the contract, trying to find more money."

To the above list, you could add scores of other examples of foolish folklore. Laugh! But don't ignore the malarkey. Just factor it into your marketing effort, especially if you know your rivals are susceptible to old myths.

SUMMING UP

Profit from your rivals' gullibility! Ignore the myths! Follow my advice and enjoy more business!

The next chapter will supply yet more help—with a list of "Fifteen Ways to Outwit Rivals."

Do's

1. Be cautious in pursuing published solicitations, especially those big ones that clients have planned for, say, the last five years.

2. Always be prepared to be a contender. Don't give up just because you think the job is wired to someone else.

3. Continually strengthen yourself—your staff, your techniques, your capabilities—if you're an incumbent. Too many people want to take your place. And some may be better qualified.

4. Propose a state of the art solution to your client's problem. No one wants otherwise. "State of the art" means exactly that—nothing either obsolete or experimental.

5. Always keep interviewing, placing ads, and otherwise looking for qualified staff whether you need them immediately or not. Amass a file of people you can instantly plug in. You might even want to hire hotshots months before you require their services.

Don'ts

1. Don't underbid a job unless you can afford it and can recover through contract expansion.

2. Do not avoid a vague solicitation because you think the buyer has wired it to a competitor. It *may* be just the product of the client's confused mind.

3. Don't let clients tag you as someone who can't follow instructions. You must follow the solitication's proposal preparation procedures exactly. Ditto for the stated requirements.

4. Avoid the belief that a rival has it all sewed up. Otherwise you'll never overcome adversity and be a true competitor.

5. Don't offer the client more than he asked for. He hasn't any way to judge the efficacy of your solution; he lacks the criteria to evaluate the unexpected. Say, you're bidding on the design and construction of a computer system to supply information on a certain process. Don't propose a powerful machine with costly, unused capabilities. Imagine using a Cray 2 to run Lotus 1–2–3.

CHAPTER 9

FIFTEEN WAYS TO OUTWIT RIVALS

You're fighting off five rivals to win a contract to design and install a giant, computerized antenna for NASA.[1]

The other bidders are offering similar prices and terms. How to shoot your rivals down?

Simple. Someone new at NASA messed up and didn't include a maintenance clause in the contract. So you say:

"Look, go with us, and we'll maintain the antenna at no charge for a year. You'll have more money left over for other projects. And on this one we'll be giving you some extra value."

No, your maintenance engineers won't work for you for free. You're gambling. But this little wrinkle could clinch the multi-million-dollar contract and open the way for more business. Over time you'll recover the costs of the engineers. Besides, even if you don't get the contract, your superior proposal will have wowed the buyers enough to increase the chance of your winning future business.

Way #1: Loss Leader—Service

The engineers will have been a clever "loss leader," a ploy to whet the customer's interest in your services by offering either a freebie or a major discount. Your customer might not even have expected to buy this little extra. That's true here. But the buy-

---

[1] A computer will help aim the antenna and communicate to (and receive signals) from the satellite.

er's ears perk up just the same. After all, you're saving NASA some money, time and embarrassment. Your expert maintenance crew can keep the antenna going no matter what glitches develop. What's more, by making the antenna more dependable, you enhance NASA's image, too—a service that you'll also render by letting the buyer gracefully recover from his failure to include the normal maintenance clause in the contract.

Congratulations if you've done something like this. You're an exception. In high tech services, too many companies lack the guts and the brains to offer terms that distinguish them from the other clones whose prices and capabilities are similar.

And that's the subject of this chapter: 15 ways to outwit your competition when (1) the other companies are as strong as you or (2) a threatening situation forces you to use offensive rather than defensive marketing tactics.[2] Please note that some of the 15 ways may not be legal to use with certain clients, government or private. Consult your lawyer if need be. You'll hardly build up good relations with bureaucrats if you encourage them to misspend tax money, even in a technical way. But ingenious bidders often can find means of offering extras while staying entirely within the law. If you can't? Then don't.

In this chapter, I'll also pass along:

- Some guidelines for developing your own tricks to win over other bidders.
- Common mistakes to avoid when you seek that extra competitive edge.

And now, having told you one of the 15 ways of outwitting your competition—by offering freebies, such as the one given NASA—I'll go on to the others:

Way #2: Other Loss Leaders

Sometimes, instead of offering a service at no charge or at a discount, you might offer a product.

---

[2] A typical example of a threatening situation might be an enemy inside the client company. Offensive tactics could include tying up your client's time so he ignores competitors. Defensive tactics could encompass a warranty on your services.

Say, you're teaching your client's people how to use electronic spreadsheets, but both machines and humans are still making costly miscalculations.[3] Then, at no charge, you might offer some auxiliary programs that verify the accuracy of the spreadsheet calculations. Your costs? Maybe $50 per copy of the audit program,

Tune your freebie to the needs of the solicitation. Avoid anything smelling of a bribe. Resist the urge to splurge on a fur coat for the buyer's wife or treat this nice couple to a trip to South America.

Rather, your offer should enhance the buyer's productivity, his position, or his authority, or reduce costs, or increase stability or security.

Way #3: The Patron Saint Gambit

Peter Jones is no saint. He's marketing director of the Sort Corporation and a hard-nosed bargainer. To the people at Beasley Bus Lines, however, Jones might as well be St. Francis of Assisi.

> Showing true concern for his customer's problems, Jones helped Beasley juggle around its funds to jack up the power of its computers and refine its software. For years the head of Beasley's data processing department, Mary Sullivan, had craved such improvements. And Sort was eager to design and install the new system. Yet it was all Sullivan could do just to afford the maintenance of the existing computers and software. She wasn't the best corporate politician. And because of the bus company's financial woes, money was tight even for the most obvious internal needs.
>
> Jones, however, found a financial patron for Sullivan who, in the end, seemed just as saintly as he.
>
> It was none other than Bill Hoffman, Beasley's budgeting director. Hoffman had attended the same college as Jones and was likewise active in the alumni organization.
>
> "Look," said Jones as the two men were whooping it up at a victory celebration of their alma mater's football team, "do you mind if I call you tomorrow? I've got a solution to your company's computer problems."

---

[3] Contrary to the dogma of some computer scientists, programs *can* err.

Jones didn't have to say any more. Everyone at Beasley knew how slow and underpowered the computer system was.

The next morning, Jones told Hoffman the plan. "You're responsible for your outfit's budget. I'm thinking you might have some ideas to help Mary get some funding to get your computer system in shape. I know you're in a crunch. And I'll do what I can to help."

"Hell, we're laying off drivers."

"Well, suppose I throw in an accounting and budgeting program that can help you do your own job better? And at the same time it'll help Mary. I'll come up with a plan to tailor everything to her needs."

"She has her pride."

"Well, maybe we can conspire on her behalf behind her back. At least at the start."

So over the next few weeks Hoffman played with his numbers and scrounged enough money off various departments' budgets to start the program to upgrade the Beasley computers. He, too, was a patron saint. Everyone came out ahead; the budget software helped Hoffman make cost cuts in the areas that were least painful. And within six months Beasley awarded Jones a contract that lasted eight years.

You can also be a patron saint in another way, by leasing equipment to someone who lacks the wherewithal to pay the price up front.

His company might make a purchase impossible—at least not without troublesome corporate bureaucracy—but he might easily be able to budget for both the lease payment and your related services. Yes, you'll have to tie in money in computer gear or in other ways. But you'll enjoy a hefty return on your investment. That's how Jones of the Sort Corporation helped market some specialized computer programs and services to managers and safety engineers at chemical plants across the country.

The software showed people how to do their jobs more safely. At $125,000 per plant, the product seemed pricey, but it might lower insurance premiums, and one manager told Jones he'd sign up— under certain conditions. The budget deadline was past. "But," the manager said, "I have this fund I can use to pay for services. Especially ones that solve problems. And that's what your software does. Now, if you lease me this equipment and the service, I won't even deplete this budgetary item. I can maintain the service every year. And I can call you back each year when I get new people or when we need refresher courses."

"Message received," said Jones, and hurried back to corporate headquarters.

Sort had to borrow several hundred thousand dollars to pay start-up costs, both equipment- and human-related. Wilthin six months after the development of the package, however, Jones signed 15 contracts to provide the computers, software, and trainers. All contracts were for five years, and some as long as ten.

Way #4: The Sage Advice Gambit

Sometimes a little free advice—to the client—will go a long way toward winning you a contract.

The Humphrey Company's customers were up in arms. Humphrey made cartons for refrigerators and other large appliances, and inevitably, just when its customers were enjoying a boom period, they'd find that their supplier's production was weeks behind. The problem was simple. Communications had broken down between sales, inventory, shipping, and other departments.

What's more, Humphrey found its cash flow problems further aggravated by a slow accounts receivable department and late mailing of invoices.

Clearly Humphrey needed a sage—ideally, an outsider, who would be more detached than someone from within. What's more, it lacked someone with all the qualifications for the task.

Adam Jennings, a marketing man with Arrow Systems, learned of Humphrey's difficulties and, at no charge, offered the services of his company. Arrow Systems might have done a simple consulting job. But they appreciated the potential for a long, lucrative contract. Jennings provided Humphrey's vice president with a long "white paper" to support the executive's belief that the company needed a more sophisticated computer system and a complete reorganization of communications procedures between divisions. The paper explored the various alternatives. Very subtly, however, it favored the solutions that Arrow was best prepared to propose in case the project went up for competitive bids. In the end, Arrow received a sole-source contract for $6 million.

Arrow isn't the only company to benefit from the Sage Advice Gambit. The Lockwood Corporation used the same ploy to win a contract worth several hundred thousand dollars from the Brown and Smith Trucking Company.

> In just a few years B&S had grown from just 3 trucks to over 60, moving frozen lettuce across the country in record time. But the company's bureaucracy somehow wasn't as fleet as the drivers themselves. Brown and Smith couldn't swiftly match customer pickup requests with trucks available at the time, delaying some lettuce shipments long enough to jeopardize the vegetables' freshness.
>
> Then Mike Ross of Lockwood Systems ran into his old bowling friend, Steve Yates, a vice president at B&S, who explained the company's woes. Ross could help. He put two of Lockwood's best people on the case: an industrial engineer and a computer specialist. They spent a week and a half reviewing trucker logs, manifests, and other paperwork, and another few days talking with some of the older drivers, asking for their solutions. The two experts then analyzed routes and truck traffic patterns.
>
> The result? B&S ended up with a new communications system and the ability to keep track of trucks' whereabouts more easily to allow for faster pickups. No more did lettuce rot. Meanwhile, Lockwood had ended up with some lettuce of its own, from a $175,000 contract to design and install the system, plus more requests in future years for maintenance and management.

Way #5: Ghostwriting

Pity this poor buyer. He might be new to his job and bewildered by strange terms such as *diodes* and *LANs* and *baud rates*.

Where to turn for help in drafting a solicitation for services or a product?

The buyer hates to reveal his ignorance to his superiors, or perhaps they are at a loss as to what to say themselves, so they're open to help from an outsider—ideally, you. Even if your game is services and he's just shopping for boxes, you might have an interest if you specialize in certain kinds of hardware. You get to know him, his problems, needs, his organization, his budgets. Then you'll help the buyer do such tasks as:

1. *Specify in the solicitation itself exactly what he wants to lease or buy, and calculate what his price range should be.* He might have to list the gear's specs. What about speed, size, power, interfaces, and compatibility (such as with IBM)? Similarly, what about schedules and quantity? Make and model? Training requirements? For an automated system, the specs might be especially complex. They may need a wide range of analytic, engineering, planning, design, development, testing, integrating and training work. Give the buyer a budgetary estimate as well.

2. *Outline what kinds of people he needs for what services.* Remember: in solicitations, people are a commodity just like equipment.

3. *Spell out the services themselves.* In detail you should describe the statement of work, special development tools, and any other special requirements. This is most critical. For architectural engineering work, the specifications could be particularly detailed, with heavy stacks of drawings. Qualifications for people and the services company might also be lengthy.

Don't forget the whole point of this exercise. You're not only rescuing the ignorant buyer and providing thorough and correct specs. You're also encouraging him to do business with your company by making the solicitation requirements consistent with what you can offer. Don't be shy. Within the limits of conscience and your client's needs, try to give your company a competitive edge if the client goes ahead with plans for public bids.

At the same time you can honestly say, "I'm not afraid of competition." And very likely, you aren't. After all, you have an inside track, a greater understanding of the client's needs than anyone else does.

To encourage the buyer to make the right choice, moreover, you can emphasize:

1. *Special design, development, or construction methods that only your company employees, or very few rivals do.* Perhaps your pet tools can reduce the number of design schedules or jack up productivity. And maybe your competitors would lack time or money to catch up with you.

2. *Special qualifications of your people.* See if you can turn your organization's staff mix to your advantage. "Qualifications" could include experience. In any event, build the qualifications using your own company as a model.

3. *Your firm's special support activities, to enhance the work done on the contract.* Maybe a special lab at your company can help your client test alternative computer memory boards and monitors. Make the test part of the contract requirements.

Of course you must be clever and cunning and make sure that (1) your selection criteria aren't that far out, (2) they pertain to the proposed project, and (3) the client doesn't think they're tuned for your firm alone.

For every unique qualification that you include, you should try to see if these are also representative of what competitors *might* have. But always preserve your competitive edge. Else, all will be for naught.

Your client's interests needn't suffer just because you've temporarily donned a buyer's hat. Consider the previous example in which Jones helped write the basic statement of work and specification for Mary Sullivan. The job was complex, and Mary was impressed. Jones's plan gave her better control over her work. And she was highly grateful.

Way #6: More Traditional Ghostwriting

Your client hates to write speeches and professional papers. The words might as well have come from the local lumber yard. And he worries—correctly—about his image.

But once more you have a solution. It's a young hotshot on your staff.

He'll ghost the speech or paper for your client. You'll add your own touches. You mustn't fail. The idea, after all, is to help the client, not hurt him. So be careful not to promise aid if it's the wrong kind.

The rewards, however, can be big. If a speech goes over well, the client might request your assistance in the future, thereby increasing his indebtedness to you. You'll be contributing directly to his professional standing. What's more, at the right time, you might even be able to sneak into the speech a flattering reference to your company—maybe not a direct one, but still one that enhances your firm's own image.

Way #7: Acting as a Surrogate at Industry Gatherings

Your client is so impressed with your company's speech-writing services that he's now open to letting you represent him at an industry association meeting or a conference.

Perhaps you'll read a speech in your client's name or describe one of your client's pet systems or techniques: ideally something in which your own company played a role.

Whatever the case, you'll be awarded two points. One will be for helping the client, and the other will be for aiding your own firm since you'll be introduced as its representative as well as the client's.

That's still another trick that Peter Jones used to steer business in the direction of the Sort Corporation.

One of his clients, Drew Oliver, an official at a public transportation agency, asked him for a favor. A respected transportation association wanted Oliver to give two speeches at the same conference, and he lacked time to write one, let alone both. "In fact," said Oliver, "I'd like you to give the second speech as my representative."

To the rescue came Sort's transportation engineers and computer people, who, aided by a silver-penned technical editor, wrote a paper on information systems for public transportation.

Eight months later Oliver released a solicitation for a system similar to what Sort's people had presented in the speech.[4] I needn't tell you who won the competitive contract. It was for $1.5 million dollars, excluding that $1 million a year for maintenance.

What's more, Sort had perked up the transportation industry's interest in the kinds of solutions that the company had to offer. Peter Jones had skillfully piggybacked on the client's credibility. The only indication of the Sort company's role was (1) The association program, which mentioned Jones's name and corporate affiliation and (2) the presentation itself, where the

---

[4] Among other things, the speech told of the transportation agency's problems and its past, present, and contemplated solutions.

name "Sort" appeared at the bottom corner of one of the papers. And remember: the program also identified him as a "special consultant and representative" for the transportation agency.

The association published both papers, including the second one given by Jones. Hence he enhanced his image along with his client's.

Way #8: The Warranty

Everything nowadays seems to come with a warranty—cars, washing machines, computers, you name it.

The idea is to assure consumers that the manufacturer stands behind the product and impress the comparison shoppers.

"Yes," the customers think, "A is not only more reliable than the competition, but A also provides a warranty against defects or poor performance. Apparently A is better built than B. Otherwise why isn't B making the claim, too? A must be a better risk."

The result? Product A's sales value and marketability sky-rocket.

Can the services business use similar techniques?

Yes, if you want to gain a competitive edge, since your competitors rarely provide warranties. But you must recognize the risk factor for your company. In the services industry, you may be the first to have put together a solution in a certain way; you aren't Maytag cranking out tens of thousands of washing machines. You lack historical reliability data or failure rate data, whether you're constructing a new kind of a house, or a computer-based system, or a new type of building energy monitoring system. What should you do?

Remember, you read earlier that you must nullify the potential of failure for the client, give him the "warm fuzzy feeling" that he can depend on you and your solution.

Well, here is the chance. Provide some sort of conditional warranty. You just don't, of course, arbitrarily spread out an across-the-board warranty. You need several conditions. For instance, the client must maintain the integrity of the software code, and can't modify it.

If the client does, you may rip up the warranty. So, to maintain the warranty, he must follow all your instructions and those of the manufacturers of any canned software you may be using. Of course the warranty can offer carrots as well as sticks. For instance, you can promise that you can incorporate any upgrades to that program. More important, you might promise to design and build the solution, that it'll work, that maintenance requirements might be low. Just be careful to weigh the risks ahead of time. Try to have some control over the operation of both the system and the installed product.

Also, you could have some price increase in your product or service to cover the cost of the warranty, but it is against certain types of failure, not an across-the-board warranty, not an unlimited warranty, not an open warranty.

You might also have separate warranties covering such issues as:

- That the solution or system can be upgraded with minimal interruption.
- Your continued upgrading will always improve efficiency and productivity.
- Your solution will include no defect to cause an interruption of the client's system or service.

We're talking about gaining a competitive edge, and not its true costs. Brace yourself for additional expenses like insurance. You need safeguards. Perhaps this means raising your prices. So be careful as to what kind of clients you offer a deluxe warranty, or any kind, period. Perhaps you'll decide to make it only an option in your proposal. Whatever the case, you must show the client that the extra costs up front will justify the long-term savings.

Ideally, however, more service companies will offer warranties of one kind or another.

Way #9: Giving to the Client's Pet Charity

Charity begins at home—the client's.

This doesn't give a big edge, but can improve relations. Many client companies or agencies have a favorite charity, whether it's the United Way, the Red Cross, the Scouts, or Toys for Tots. Help your client make its yearly quota. It'll look good. So will your company. You'll feel better, too.

Way #10: The Free Lunch

Most clients really like to be taken to lunch and even to dinner. Even those wary of free gifts and other expensive favors will leap at the chance.

In fact, most prospects think that something is wrong with you if you *don't* take them out for a good steak.

Free or not, a lunch is a good escape from the office, from interfering phone calls, and from other people interrupting. It's a way to swap information. Your client will know you faster, and vice versa. Yes, his boss may ban free meals, but the client might still like to go out for a Dutch treat lunch. Plus, studies show that prospects are more susceptible to pitches over food. Just make sure that neither your pitches nor the steaks are overdone. Remember that you're a marketer, not a huckster.

Also be low-key as to who pays. If the buyer's organization bans free lunches, you might say: "Hey, it's okay. I'll pay this time. The next lunch is on you."

This, of course, sets up an invitation for another lunch.

Trade shows are a special opportunity for free-lunchers—both the givers and the receivers. Feeding prospective clients won't win you a competitive edge. But they'll gain you new contacts. So within the bounds of your budget—remember, trade shows are not your main marketing focus—see if you can set up that hospitality suite.

Other lunch opportunities abound. You don't even need a trade show as an excuse. For instance, you can set up a seminar on special techniques, solutions, approaches, and planning methods, or something about the way you do business that is special. Tour the country. Rent hotel rooms. Offer a standard program. Send out invitations to prospective clients in the areas you're passing through. Give them a speech on your techniques, or your experiences, and your qualifications at no charge. Fill up your prospects with free coffee, cakes, and sandwiches.[5] What a wonderful way to promote your company.

---

[5] Or charge a $10 fee if your budget calls for one and you think your message is popular enough.

Dinner parties are much more serious than lunch—especially those given at your house, since the expense is higher. But of course after hours you can invite other guests, maybe even some VIPs to impress your prospective buyer. You can also include top executives from your firm and the client's, and, obviously, the spouses. I owe my career in no small part to Sandra's cooking. Not to be sexist; if I were the better cook, the reverse might be true. Whoever is doing the entertaining—wife or husband—keep this in mind: Spouses play an important role, as clients' spouses will often learn more about your own character, accomplishments, stability, and adherence to business and societal norms through conversations with your wife or husband and visits to your home. The conversations will reach your client—and serve as sort of a reference check.

Granted, business dinner parties can be a bit "tribal," but, nonetheless, they can engender rapport and trust. And you need not be bashful about their purpose—business. After all, you've brought together all the principals from your firm and from the client's organization.

Way #11: Being a Joiner

Clients and buyers love to recruit for professional societies and trade associations. Join! Move up in the ranks. Organize parties. Start and run subcommittees. Do the grunt work. Gain influence. Become an official of the association. Get to sit on the dais at the annual dinner.

Then reap the benefits.

Bring along your company's senior professionals to meet prospective clients at the association's dinners and luncheons.

Your organizational work won't just impress the prospects. It'll also win you a wealth of contacts and influence; you may even play a role in the drafting of industry standards and procedures. You may also be able to appoint people from your firm to serve on the key committees and present major papers and make contacts of their own.

Such activities won't give you a definite edge on a specific competitive bid unless you are actually working with a prospective client through the professional organization. But they're a wonderful long-term investment. Now you have a legitimate reason, particularly in government marketing, to deal with the top generals, admirals, navy captains, and GS–15s and 16s at a peer level.

Having joined professional societies and trade associations, you can also participate in trade shows and professional conferences to demonstrate that you mean well and that you are serious about participating. That's true even if you have to set up a booth just to market a service. The association will appreciate your support. Peter Jones of Sort used this booth strategy to increase his standing and visibility within public transportation groups.

Mixing with members, presenting papers, setting up booths, and otherwise acting as a true peer, you'll find that your name and your firm's will pop up again and again in newsletters and professional journals. You're no longer a stranger. People will regard you as a known commodity—and a likable, public-spirited one at that.

This is the way to build your image up, not through any single booth at a single conference. And you'll do the same for your colleagues and company. You'll all permeate the professional organization; you'll support it; you'll finance it; you'll grow with it.

Never forget that the importance of the association is important to members of the client industry. Otherwise they wouldn't be participating; they wouldn't be showing up every year at the annual conference.

Eventually, believe me, this investment will win you a competitive edge, not only with a single client, but with a host of prospective clients. What's more, the good professionals on your staff will appreciate your efforts. Groups can contribute wonderfully to their personal contacts and knowledge. And your blessing will make the experience still more enjoyable.

You of course run the risk of rivals pirating your people because they're more visible. Then again, if your company is growing and you're letting your people grow professionally, the dangers are much less than you'd expect. Tell your people, "Go ahead and share the professional limelight. Our company *wants* leaders." You may then enjoy some defections in the other direction. The first-rate professionals at other companies will be envious of the freedom you give your performers.

Way #12: Stewardship

Ken Turner was one of Peter Jones's best clients. Turner was the data processing manager for one of the bigger hamburger chains in America, a sprawling corporation where bureaucratic factions were constantly trying to make mincemeat of each other. Thanks partly to Jones, however, Turner not only survived, he thrived. Jones and the Sort Corporation's other marketing staffers, knowing who buttered their bread, were constantly alert for important information that Turner could use. If they heard a regional manager grumbling about the response time of the chain's mainframe, they'd pass this information on to Turner, who'd then be prepared when the complaint officially reached his desk.

The fatherly Jones also counselled the young Turner on how to fight his bureaucratic battles. And when he played golf with the chain's president, guess whose name often came up in a positive way?

In short, Jones was acting as a steward—constantly alert for ways to protect and advance his client's career.

Appreciating the gossip and hard information that was coming his way, Turner pressed aggressively for an expansion of the Sort Corporation's services contracts. After all, the more Sorters he employed, the more "ears" he'd enjoy.

Sort's stewardship tactics, of course, wouldn't have been enough to compensate for bad work. And imagine the opportunities for disaster if Jones and his people hadn't been true marketing diplomats who could get along not only with Turner but also with his rivals. They also were good secret keepers. Furthermore, despite Jones's fatherly advice, he let Turner himself call the shots. The Sorters were there to help the client, not boss him around.

Forget about the stewardship approach if you—or your people—lack the savoir faire of the Sorters. If you possess it, however, the financial rewards can be big.

Way #13: A Siege Campaign

Ann Bosworth, a crack marketer for Midwest Information Services, badly wanted a contract with the Kowen Company, a large manufacturer of farm machinery.

Her colleagues, however, liked to joke that she was on "Mission Impossible."

At least half a dozen rivals would be competing for a $7 million contract to design and build Kowen's international communications system. All did good work for low prices. On top of that, Ann had a special problem. Her ex-husband, whom she'd divorced years ago, and with whom she was still on bad terms, was one of Kowen's executive vice presidents. In other words, she had an enemy within the client organization.

Bosworth responded with both strong defenses and a powerful attack—as if she were laying siege to a whole city of warriors.

She cranked out position paper after position paper for her few sympathizers within Kowen, attended luncheons where she could win new friends, and stepped up her already heavy participation in the professional association where she knew Kowen's data processing people. She made herself too helpful, too indispensable, to be ignored. And despite some original misgivings— partly the result of poison spread by Bosworth's ex-husband— the Kowen people found her solutions irresistible. In the end, her diligence and the quality of the work proved too much for even formidable foes. Time after time they would call, only to find that the DP director was tied up in a meeting; and very likely it would be with the ever-present Bosworth or her technical staffers.

Kowen's confidence in her steadily built. In an amazingly short time, DP people revealed sensitive information—such as plans, schedules, and budgets—that they never shared with Bosworth's rivals. Hence, she could come up with a more responsive offer. Bosworth knew all the effort was worth it. The very fact that Kowen shared its secrets shows that through her intelligence and hard work she had become part of the family. She was inside the walls. Her competitors were still outside.

Few marketing people, alas, will mount a Bosworth-style

siege. All too often, buyers (in government and industry) complain that bidders don't follow up with enough calls on good prospects. What a pity. The buyers *want* to hear more—and, often, want to communicate more about their requirements. And so these clients lose out on a chance for the most responsive bids, and good services companies fail to win contracts.

Way #14: Becoming the Incumbent

Want to win badly? Want additional contract work? Here's one of the best ways of all: Become the incumbent contractor.

That sounds like a catch-22 but actually isn't. Just remember that you might be better off giving the client a break in the beginning in hopes of picking up more business later on. Such a strategy of course will require:

- Top performance.
- Good rapport with the client.
- A campaign against your competitors. The main idea isn't to bad-mouth them but to obligate the client to you so much that you've shut out your rivals, while you're building up your company, its reputation, and client relations.

One beauty of being an incumbent is that your marketing costs will be minuscule compared to those associated with a new client.

Way #15: Showing Sensitivity

Be more sensitive to client's needs than your rivals are. For instance:

- Be more advanced than state of the art—through a first-class research and development program.
- Give your professionals the special tools they need to improve productivity without stress.

In short, build a formidable, healthy, thriving company where managers, professionals, and administrators alike are gung ho and the clients can appreciate this. Cut down turnover.

What better way to show clients that you can nullify the potential for failure within your own company?

A strong company with enthusiastic people, moreover, is the ultimate protection against Murphy's Law. Say, a key proposal writer gets sick just before the deadline. Then his colleagues can instantly leap into action to finish the job. Clients have mysterious ways of knowing that if professionals are sensitive to each other's needs for reliability, they'll be more sensitive to outsider's needs as well.

GUIDELINES FOR DEVELOPING YOUR OWN SPECIAL WAYS OF WINNING OVER RIVALS

In coming up with your own special tricks, keep in mind some of your client's big priorities—for instance, saving money and time and increasing the reliability or availability of crucial equipment.

You must deal with (1) "hot buttons" or special buying reasons and (2) latent drivers like client pride and fear.

Hot Buttons

Here are four hot buttons:

1. *Dollar savings.* Your client wants to save money directly if possible, not just over the life of a computer or other piece of equipment. He'd love to have spare money that his budget normally wouldn't allow.

2. *Time savings.* He wants a more productive staff; what better way to enhance his standing in the company? And if he can save money indirectly in this manner—well, so much the better.

3. *Increased control.* The client, his company, or a combination of the two wants increased control over processes, information, or people. Take a hamburger chain. If its headquarters can enjoy daily sales reports from every location, the executives will feel more in control than if the information came only monthly. They'll be better able to provide guidance for their regional people and perhaps even for the individual managers.

4. *A general and comprehensive category of reasons that I call the "* . . . *ilities."* They'd include:

Reliability.
Maintainability.
Controllability.
Operability.
Supportability.
Transportability.
Portability.
Expandability.
Flexibility.
Integrity.
Safety. (More than an honorary "ility.")
Security. (The same.)

Weaknesses or faults in those areas could show the need for new or added-on work. Respond accordingly! Help your client see gaps that he doesn't perceive in the company operations. Perhaps they require the equivalent of a Chevrolet sedan but he really wants a Rolls Royce. Don't sell him a Chevy at a Rolls price. But try to address the weaknesses of the Chevy. Maybe the client worries it'll break down more often than the Rolls; so you might want to throw in, at no charge, the equivalent of a tool kit.

You must carefully probe the client to see if the tool kit or other extra will have its desired effect. Don't surprise your client. See if he'd really appreciate your extra effort.

Latent Drivers

Latent drivers include client image, pride, fears, hatred, competitiveness, greed, and curiosity, among other feelings and compulsions.

Watch for anything falling in those categories. Say, your client's boss might let him buy a computer that can store 500 megabytes. The prospective client has a corporate rival, who's also hoping to get a computer system; in fact, maybe the boss will allow just one. The rivalry heats up. And ideally you'll figure out a way to help your client buy a 750-megabyte machine for not much more than the 500-megabyte model—to show that he's a better spender of company funds than the other person is. Yes, you'll help your client do his job better. But you'll also be getting your hooks into him. He'll feel that he can't do without your services, that he must choose you over the other bidders.

If you know of latent drivers—such as the rivalry that made the buyer eager to outcompute the other guy—you'll really have a chance to shine even if a hot button isn't begging to be pushed. Most clients' solicitations don't include all their *true* requirements, especially emotional ones. And you can take advantage of such a gap.

MISTAKES TO AVOID WHEN SEEKING A COMPETITIVE EDGE

Below are some actions that usually do *not* offer you a competitive edge. Note that many of them are common responses of companies selling computer products. But you're different. You're in professional services.

The actions:

Price Cutting

You might cut your prices on a single bid to gain what you think is some degree of a competitive edge. But you cannot continue to operate under this kind of a price regime.

What you must do is, in essence, permanently cut your costs of doing business. In some cases, cost cutting may win you the job—you might successfully buy in. But drastic price cuts on bid after bid will kill your company. Only big, deep-pocketed firms can play this game. And even they can experience uneven results. So limit your cost cutting only to jobs where the odds are good for expansion that will let you recoup your costs.

Increasing Channels of Distribution

You could set up field offices to be near your major clients, assuming also that there is a labor supply available in that community. But think hard before doing so. A much cheaper solution could be express delivery services and low-cost phone lines.

Unless you're large, don't mess with more than one or two field offices.

Increased Advertising

Only in recent years have professional services firms engaged in heavy advertising—for example, attorneys or health maintenance organizations (HMOs). Keep in mind, however, that the lawyers and HMOs are going after consumers. Engineering services companies, haven't used advertising successfully except to polish corporate images. What's more, if you're medium-sized, advertising normally won't work unless you keep the ads on the air or in print for a year or two or maybe even longer.

Buyers won't rush out to buy services, even when your ads bombard them, unless they need your company. You aren't Procter and Gamble; you aren't advertising a product, such as soap, that readers and reviewers will need steadily.

Of course, some particular type of high tech service might be in such great demand that ads will pay. But I don't know of anyone trying this on a large-scale basis. Why? Perhaps because high tech services companies lack the money for such a grand experiment.

In much the same way, trade shows won't help you except by building goodwill and by helping you gather intelligence on rivals. Sorry.

Maybe some marketing genius will come out eventually with a new approach guaranteed to bring in customers through ads and trade shows. If so, I wish him luck. He'll need it.

Improving Packaging in the Wrong Ways
(Although There Are Good Ones)

Here again, you aren't P&G trying to sell more soap by changing the color of the boxes. There are only so many ways in which you can improve the packaging; don't preoccupy yourself with trifles such as whether you send out your proposals with green or red covers.

You can, however, guard against typographical errors and use professional-looking formats.

You can also encourage your people—the ones in contact with customers—to dress professionally. Beyond that, you can keep your corporate headquarters in good shape and locate it in a reasonably impressive building if your budget allows. Why? Because clients expect businesses and business people to reflect a certain image. Sloppy and cheap looking offices without standard status identifiers—for example, corner offices and carpets and plants for executives—mean to many clients that the company is sloppy and cheap. "Ah-ha," they think, "those bozos don't even confer proper status on their executives." Yes, I know: Some of the so-called '60s Generation will laugh at this emphasis on externals. If so, don't blame me. I'm just telling you the way things are in high tech services.

Another form of packaging? Good business ethics. I'm not slighting them by including them this far down on the list. Indeed, "Honesty" is part of the Friedman Formula for Competitive Success. I'm just reluctant to include ethics in the "packaging" category. Ethics should be acted out, not be mere "packaging." Of course, a pile of clippings about SEC indictments won't do wonders for your company's image.

The above are about the only ways that you can package a high tech services firm. And for better or worse, the results won't be as dramatic as the benefits of new soap boxes.

Bad-Mouthing Competitors

You won't win by bad-mouthing Company X.

Buyers are usually suspicious of companies that put down their competitors for several reasons.

For one thing, buyers feel that the bad-mouther is really trying to hide some of his own weaknesses. And they are also afraid that if you say nasty things about competitors, maybe some time in the future you may be saying nasty things about your clients.

Buyers just don't trust people with bad mouths. So don't go around saying that the competitor is a liar, or a cheat, or a child abuser.

Now, you might be able to say or hint that your competitor just had his last contract cancelled in default; just make sure that's both true and apropos. But buyers normally have an inside track on such goodies and learned of them long before you did. So watch that mouth. Avoid nasty remarks about competitors' past performance or proposed solutions. High tech services isn't the fast food business. You're not on TV saying, "Brand X denies you freedom of choice in hamburgers." Or, "Brand X flavors its meat with dioxin."

That's overstating the case. But you get the point? Just because your clients may attack their competitors, you needn't attack yours.

Granted, you occasionally can attack a competitor to show that you and your solution are less of a risk—a topic discussed in the chapter on proposal development. But that's the exception. The client or buyer isn't expecting you to attack your competitor and may even resent you for it.

Other Don'ts

Now, here is a short list of other don'ts—some of which might even cause you to lose a contract. In this sense, these are the opposite of gaining a competitive edge. The don'ts:

1. *Don't hire your client's best managers or employees.* Do not pursue your client's favorite people unless, of course, they are dissatisfied and come to you looking for a position.

2. *Don't shower gifts all the time on your client and his staff, especially gifts like a new car or a new home.* Maybe next year's calendars are just right. Avoid letting improprieties—or an appearance of them—raise questions about the quality of your services.

3. *Don't hide or lie about contract problems.* Don't surprise your client.

4. *Don't side with your client's opponents in the firm and don't overrun his budget, at least, more than once.* Why jeopardize your chances of winning another contract or an expansion of the existing one?

SUMMING UP

There are many ways to gain a competitive edge. Probably you'll have to change each tactic for each bid, or use a combination of tactics for each bid. Buyers are different for each bid; buying reasons differ, too; so do fears and desires. If, however, you're sensitive to buyers' individual needs, you can indeed gain a big competitive edge. Be imaginative. Break out of the usual marketing modes. Don't be just another clone company.

And don't restrict your imagination to the solutions yourself. See if you can show just as much flair in the actual planning and writing of your proposals.

Do's

1. Take calculated risks in making front-end investments—such as in machines or testing facilities—if you want to gain a competitive edge.

2. List no-charge items under standard accounting practices.

3. Put no charge on labor categories that you probably won't use. Say, the solicitation requests a junior programmer but you think the work won't require one. Then you bid with no charge in that category, taking your chances.

4. Provide quality services at a reasonable rate, backed by a warranty.

5. Search the solicitation for possible missing services—services traditionally provided—then offer them as options. As mentioned earlier in this book, however, don't impose unwanted, unneeded services that clutter up the selection process.

Don'ts

1. Don't throw in a no-charge item for its own sake; make sure you can justify it as part of your capture strategy.

2. Avoid being just a bystander when you join a client's professional organizations. Participate in all the organizational functions.

3. Do not embarrass your client by offering badly researched position papers. Otherwise they'll harm you. Just one slip-up can screw up your relations with clients for the next five years.

(See other Don'ts beginning on page 181.)

CHAPTER 10

THE 100-POUND SALES BROCHURE: HOW TO DEVELOP A WINNING PROPOSAL

The Cort Corporation beat three competitors, including its arch-rival, Qwerty Systems, to win a $50 million contract to build a national network for P&F Tax Miser Inc.[1] The network would transmit work from local P&F offices to corporate headquarters in Miami or to regional computer centers, which would help check the processed tax returns and monitor and shift the work flow. The contract included not only the network itself but also management, operation, and maintenance of the dozen computer centers. Cort would furnish tens of millions of dollars worth of computers, ranging from AT-class micros in local offices to gigantic mainframes for the regional and headquarters offices.

All the competitors were top flight. Cort, however, offered something extra to entice the P&F vice president for operations—a 100-pound sales brochure. That's indeed how much the proposal weighed, the same as most of the rivals'. But this one was different. The 100 pounds said, "Buy me, buy me!" Rival proposals just droned on without focusing on the problems that P&F needed to conquer.

Cort, on the other hand, keyed its 10,000-page proposal to the client's solicitation—not in a vague, halfhearted way, but item by item.

---

[1] For clarity's sake, this is a composite case example.

The marketers and proposal writers examined P&F's buying reasons. And they didn't just go by the specs. They also based the proposal on the hundreds of hours they'd spent with P&F beforehand. Cort had geared the whole marketing process toward satisfying the buyer: everything from lead generation, to market intelligence efforts that uncovered weaknesses of rivals, to the dozens of lunches with P&F's people where Cort successfully promoted its solution. So the buyers ended up reading exactly what they wanted.

Furthermore, while writing the 100-pound sales brochure, Cort's people took pains to protect themselves.

An assumptions and constraints section, for instance, sharply reduced the chances of the Cort Corporation and P&F squaring off in court over an alleged cost overrun. P&F, in fact, had actually favored the section to help make the contract terms more precise.

If you want to be another Cort, you should learn:

1. *Common mistakes that Cort's proposal avoided.* Some bidders, for instance, write their proposals as if they're dashing off the script for a TV commercial. You have more than a minute or 30 words to make your case; so avoid corny, idiotic, unsupported boasts.

2. *The advice of Jerry Kenner, Cort's ace marketer, on how to design a proposal that responds to your client's needs better than the competitor's bid.* If you have only 30 days in which to develop and deliver a proposal, you should spend at least half that time planning the proposal and preparing its design. Whatever your plan, don't forget your big goal—not just to show you meet your client's needs, but to do this better than any other company in the world. It isn't enough to meet requirements. In each area that the proposals cover, your buyer will compare rivals to what he perceives as the strongest offer in this respect. You might say he's grading on the curve—reflecting standards set by the strong bidder.

3. *What a typical proposal should include.* I won't discuss here all the mechanics of organizing a proposal—for instance, whether to use a story board or shuffle around file cards. Nor will I go into every section of Cort's proposal; contents will vary anyway, case by case. And I won't waste time telling how to write the perfect sentence or perfect paragraph.[2]

Rather I'll tell what a client expects to read in the main elements of a proposal, such as the letter of transmittal and the executive summary. Usually the sections covered here show up even in small, $1 million contracts that run only 100 pages.

Whether you're directing the writing of a 100-pound monster for Cort or are chasing business for a minnow-sized company, the same ideas would apply.

And foremost of these is the need never to forget client requirements, stated and unstated. Too many services firms submit slapdash proposals that don't address the big buying reasons.

No matter the omissions, you must be wary of developing proposals that are doomed to failure. You may very well address client requirements, but then your message might not get through, and the client may not be able to justify his selection of your firm.

Below are four types of proposals to avoid.

The *Pate de Fois Gras* Proposal and Other Losers

In France, farmers grab a goose by the neck, open its bill, put a funnel in its gullet, and pour down corn. Gourmets eat the results: *pate de fois gras*. Corn-stuffed proposals, however, aren't nearly as appetizing. So avoid pouring on the corn. You aren't writing a TV commercial. You're going for the brain, not the gut, not the glands; your readers won't hear the "Noxema Girl" sexily cooing your message. You can move mountains of skin cream that way—but not high tech services.

So don't just say, "We are the largest company in this field." You must follow with explanations of what makes you the largest firm—more contracts, highest sales, largest staff in the

---

[2] Just read *The Elements of Style,* by William Strunk, Jr., the mentor of E. B. White, the famous essayist. Published by the Macmillan Company in New York City, the book is dear to English professors everywhere, or at least the better ones.

country doing this work. Then lay out the evidence for the claim, such as a count of all the contracts completed.

And now the other losers . . .

The Bag of Worms Proposal

A Bag of Worms Proposal is one in which design plans are chaotically mixed with work plans and management plans in the same section.

Don't write this loser. Avoid a disjointed proposal that reads as though a committee dashed it out, trying to pack in everyone's pet idea. Don't just throw in a worm bag. Organize! Be focused!

For example, don't let the same section include both (1) your design approach and (2) your work and management plans. The evaluator will hate you. He doesn't want to have to cull out the design material from the rest. So write your proposal according to major themes and objectives, and remain consistent within each section.

The Lick and Promise Proposal

You lack time to read the solicitation. You were short staffed. And you haven't filled in the gaps. You have not responded completely to all the requirements of the RFP. In other words, you're ready to perpetrate a "lick and promise" proposal. Don't.

Instead, respond completely to all requirements—specifications, statement of work, contract terms and conditions. Don't just say, "We will provide equipment that meets your specifications." Nor should you merely say, "We will do all the work identified in your statement of work," and then list all the tasks.

Also, don't just brag, "We have conducted over 100 computer security accreditations in the last four years." Rather than just trading on your company's reputation and track record with foolish generalities, you should precisely tell:

- What your company will do for your client.
- How it will meet statement of work task requirements.

If you can relate your track record to the specific tasks that the solicitation wants—well, so much the better. But don't force the client to guess whether your past experience will help you meet his *exact* needs.

The Parrot Proposal

Only the most unqualified of contractors will write a parrot proposal. It just repeats the statement of work tasks and says, "We will do this."

Government agencies automatically reject parrot proposals. Private purchasing agents at the very least find them to be embarrassments—not just to the services company but also to them. Who would ever want to attract such air-headed bidders?

Your proposal writers should be exactly that—creative *writers,* not parrot-typists.

HOW CORT DESIGNED A RESPONSIVE PROPOSAL

Jerry Kenner, Cort's new marketing man, was an old pro. You name the high points of an era, and Kenner was in the thick of things.

In the mid-'60s, for instance, he'd packaged some computer-related proposals for the War on Poverty. When the oil companies were soliciting services for the Alaskan pipeline, there was Kenner—marketing for his employer. He had one of the best win rates in high tech services. And that was why Cort, suffering a steep decline in recent years, had stolen him from a competitor.

Kenner had leaped at the chance. He worried he was growing stale in his old job, and he loved Cort's big specialties—computers and communications. In fact, Kenner was the first person in his neighborhood to buy an IBM PC, with which he wired in memos to his office via the phone lines and occasionally logged on to one of the more staid BBSs[3] around town. At the

---

[3] A BBS is an electronic bulletin board system on which computer buffs leave and pick up messages over the phone lines. Typical topics are technical.

same time, like most high tech marketers, he was hardly a computer nerd. He was a Big Brother, for instance, and active in church and civic groups. And his executive style reflected his empathy for others—one reason, beyond his considerable technical skills as a market engineer, that Cort had hired him.

During his first week on the job at Cort, the company asked him to give a pep talk both to a small group of key proposal writers and technical people and explain the principles of good proposals.

"First thing you want to remember," said Kenner, "is what drives our proposal. And that's our marketing intelligence. And our client liaison. In the case of the P&F contract we've been telling them we'll get their system running faster than our competitors will. And we'd better not forget that when we do the proposal itself.

"The marketing intelligence we have gathered, plus the official solicitation, will drive our technical strategy. Our contractual strategy. Our pricing. Our staffing. Our management and project control approaches. Our work breakdown structure.

"So we'd better design our proposal carefully. Remember, P&F will probably make it part of the actual contract. Which means it'll influence how we do the job, and how we guard against cost overruns. You've got to think of everything. The scope of the work. What services P&F needs. What its problem areas are. Like the line noise they complained about between Miami and Philly in their current system. Or the fact that they need the new system up and running yesterday. In short we're going to tear P&F's solicitation apart, sentence by sentence, and respond absolutely to each requirement. Questions?"

A senior engineer spoke up. "What if we can think of better ways of doing things than what they have in mind?"

"The proposal *must* respond to the client's wishes," said Kenner. "Write a response matrix.[4] It'll assure both you and the client that you've addressed and responded to all the requirements. And it'll inform him where he can find our responses in the proposal. If you have a better way of doing things, wait until

---

[4] To be discussed on page 202.

the contract to discuss it with him. Of course ideally your client is already familiar with your pet solution long before this. You think we go to lunches with clients just to be gourmets?"

"Well," asked a young proposal writer, "what are our big, operative concepts?"

"That's all you're asking?"

"Yes."

"I can tell you've worked in government," Kenner said to laughter.

Then he listed the eight concepts explained below:

Concept #1: Decide what the client must know and be convinced of.

Concept #2: Figure out how you'll show you will reduce the client's risks.

Concept #3: Plan how you'll prove your qualifications.

Concept #4: Decide if you *really* can do the job.

Concept #5: Review the solicitation's contents requirement by requirement. You're now thinking at the micro level, in details.

Concept #6: Make good use of marketing intelligence on the client and your rivals.

Concept #7: Develop proposal strategies. For instance, they might tell why the client will come out ahead using *you*.

Concept #8: Organize the best proposal team.

Concept #1: Decide What the Client Must Know

"First," said Kenner, "the proposal will include our approach to solving the client's problem. Which is to get a reliable data net and related computer facilities going in time for P&F's big tax season. We'll use our own regional computer centers.

"Second—we'll tell what we'll guarantee as a result of our services. Namely, a dependable net with enough capacity. And high productivity. So they can process more tax returns, faster. And more accurately and cheaply.

"Third—let's give the specific details of what we're proposing. And how we're proposing it. And our qualifications, beyond Cort's having been astute enough to hire me. You're the best, too," said Kenner, looking at the engineers, "which is why I came to work here, and why we're going to win this proposal. And we're going to remind P&F of this in a way that they'll believe."

Concept #2: Figure Out How You'll Show You Will Reduce the Client's Risks

"P&F must think we're the safest outfit to deal with. Barring nuclear war, their net can't fail. We'll build all kinds of redundancy in. And we'll let P&F know about this. In a believable way.

"We'll also do a little prototyping in our lab. We'll check out the communications program we're thinking about using. We won't give it a full test. But we'll see how it works over the phone lines when we're transmitting data of the kind that P&F would have in mind. We'll prototype and test out some of the applications software. And we'll make sure that unskilled people can operate it well. We'll turn a few secretaries loose on it. Or maybe some people recruited from some temporary help agency. We want our system to be failure-proof. And productive. We'll time how long it might take to generate a tax form on each microcomputer. If need be, we'll increase the number of machines in each local office. Even if we can't test ours in full, we can at least try out some parts and extrapolate to the whole. And then we'll let P&F know about this.

"Above all we'll show our professionalism with a good, well-organized plan. And we'll also prove that we can reduce other risks. That we can make their deadlines. That we're a solid company financially."

Concept #3: Plan How You'll Prove Your Qualifications

"Actually P&F already knows in general how great we are. Now we've got to show how we're great for this particular job. We won't be Muhammad Ali screaming, 'I'm the greatest.' There are other ways. Like being specific. You can't just say 'the greatest.' You've got to say, 'I'm good enough to have KOd Joe Blow with a right hook. Even though his reach was 4 inches longer.' People, we've got some of best CPAs and communications experts in the country. And the best computer services bureaus around. That's our right hook. I was looking at a list of our people's technical papers in the communications area. Very impressive. And I know we've got some ace programmers with experience in tax preparation software. We'll have 'em rework their resumes to emphasize this.

"Of course P&F will value good references that our clients have given us as a company. They want to know we're strong in the field. But let's not overwhelm them with our track record. They're not buying our past solutions for others. They're interested in our solutions for *them*.

"I want our technical people to keep talking to P&F's and find out their needs. So our proposal can reflect this. We've got to show we have a sound plan that'll work for P&F."

Concept #4: Decide if You *Really* Can Do the Job

"First off, let's review our own bid and no-bid criteria. Let's make sure we want the job as badly as we've assured the P&F people. Let's identify our strengths. Our weakness. And how we can overcome them. For instance, our documentation. Some of it is gobbledygook. And we're going to have to shape up. Because we'll design the system for those part-timers working for P&F Tax Miser. And for P&F's managers. The end users. You think P&F hires them with a Ph.D. in cybernetics? Above all, we've got to make sure our system satisfies P&F's clients with accuracy and speed."

Concept #5: Review the Solicitation's Contents Requirement by Requirement

"We'll focus on the solicitation as we've discussed—sentence by sentence, requirement by requirement. We'll create a checklist. Like I've said, we'll recognize the various subtleties. For instance, if a P&F requirement says *shall*, we know it's a *must*. A *should* allows a little flexibility. Same with *will*. That's the way the government often works things, and many private markets think the same way.

"Most of all, we'll pay attention to the solicitation's proposal evaluation criteria. It's key; the way to determine the focus of our responses to the solicitation."

The General Scope of the Work

"To analyze the scope of the work, we'll review the scope of each task and the extent of contractor responsibilities and efforts to be performed. What does the client want us to do or deliver for each task?

"This information will feed into our technical approach, our list of deliverable products, and our management and work plans.

"Let's pay attention to the extent to which we'll take over P&F's main database needs. That could be a burning issue later on. We're helping 'em switch work around, but the solicitation's a little unclear about our responsibilities vis-a-vis the main database operation.

"With the general scope analyzed, we'll then work up charts showing our proposed program. What we'll do. How we'll manage it. How we'll organize the work flow. We've got to show we can manage the program."

Concept #6: Make Good Use of Marketing Intelligence on the Client and Your Rivals

"We can use market intelligence to place the solicitation's contents in perspective. What you see in print isn't going to tell everything. Our clients may have needs they simply haven't thought of articulating in the proposal. And I'm depending on our technical people to come up with the little scraps of information that the solicitation might not have.

"For instance, we already know about that Miami-Philly line noise problem they're having with their existing communications. But "—Kenner held up a copy of the solicitation—" you don't see a syllable about it in here. In other words, people, think behind the solicitation itself when you do your analysis.

"And also think about our rivals while you're analyzing.

"People, we've got some good information about Qwerty. Seems they have just about all their computing resources tied up. And no regional centers. Just headquarters. Which means they'll depend much more heavily on Tymnet and service bureaus than otherwise. We can squash 'em on costs and pricing—both!"

Concept #7: Develop Proposal Strategies

"People, we've got to decide what our big strategies and themes will be in the important areas.
"Here's my thinking right now."

Capture Strategy
"What are P&F's big buying reasons and how are we going to solve their problems?
"Simple. Lower cost per tax return. Faster tax preparation, and more accuracy, too. Ability to handle tax forms for all states and major cities. Improved management controls. Improved accountability—so customers can know who made out 'this damn tax return.'
"We've got three parts in our capture strategy to respond to the buying reasons:

1. What we do better than anyone else. Our computer centers. Which we'll use so P&F won't have spent millions on its own computers.
2. Use of our automated tax income models. They'll improve tax preparation productivity and throughput.
3. Use of our automated audit programs. We'll help P&F verify accuracy of the final tax form. And offer traceability. And assignment of accountability.

"These capture strategies will become part of the technical strategy."

Technical Strategy
"P&F will now have a computer system that will be available two thirds sooner than if they bought and installed the equipment themselves.
"And we'll have field-tested their software. And we'll build in some rapid-tuning ability. To accommodate laws and tax changes at the state and local levels.

"What's more, we'll supply our own AT clones—already tested and ready to install at P&F field offices. And we'll maintain them.

"We'll give 'em one-stop shopping. Everything they need. Equipment and management from a single source. They won't have to buy or rent that much equipment."

Cost and Pricing Strategy

"We'll help P&F bottom-line it. We'll cut their DP and communications costs to the bone. We aren't just good. We're cheap. And there's a nice surprise, people. We'll probably be low in regard to our competitors. But not by our own standards. Let's just say that cost can drive technical strategy and vice versa. Here our technical strategy's letting us offer the best. By using our computer centers and network, P&F will only pay for actual use of equipment and software with heaviest load during tax preparation season. During off-season there will be little use and little expense. If P&F builds their own centers and buys their own equipment for the full network, most equipment will be idle much of the rest of the year. We'll price the job based on actual computer usage."

The Main Message(s), the Compelling Reason(s), to Choose You

"Well, you already know about our price and technical advantage. And that's not even to mention our tax experience. But our competitors very likely will make the same point. I'd like to go *beyond* them. Let's focus on where we're unique. Why, we've set up automated tax systems from Maine to Hawaii. All 50 states. And we've supported tax centers in 37 states. Plus, we can reduce P&F Tax Miser's competition. We've already drafted tentative agreements with independent tax preparers to use the Tax Miser work stations. And the software and the network. So an independent in Illinois can find out the tax laws in New York. Why, Tax Miser can even bring the independents into their organization. And even if that doesn't work out, they can collect some fat fees for software and network use."

Proposal Format and Design

"You've got to design before you write. If you don't, you'll be like a builder putting up a skyscraper—without having blueprints in front of him. Another parallel: Imagine you're writing specs for the development of a piece of equipment or a computer program.

"Design well, and you'll provide the detailed guidelines needed to write the proposal. The writer will know exactly what to write. Then he'll integrate all proposal material. He'll cover all strategies and requirements. And he won't deviate from major themes. If you skimp on proposal planning, you will only have to make it up during proposal development—meaning extensive redevelopment and rewrite.

"The following are principal guidelines for designing the proposal:

1. Follow whatever the proposal format instructions are in the solicitation—tuned according to the evaluation criteria.
2. Develop a topical outline, by sections and subsections.
3. Identify the subject matter of each section and subsection.
4. Relate the solicitation's requirements, statement of problems, specifications, statement of work and evaluation criteria to appropriate sections and subsections.
5. Develop the response matrix, which is sometimes called the compliance matrix.
6. Identify the major strategies and themes for each section and subsection.
7. Develop, for each section and subsection, detailed descriptions of what has to be written: an outline of each paragraph, with its main subject. Indicate the requirements that must be addressed, and graphics to be included.
8. Identify for each section planned page count, assigned writer(s), and schedule for each draft milestone.
9. Identify all resources needed to develop the proposal, including staffing—writers, cost people, editors, word processors. And schedules. And levels of effort.

"Now I'll identify the major members of the proposal team."

Concept #8: Organize the Best Proposal Team

"OK, people, so you'll know what's happening, here's how I've selected the proposal team. This will give you an idea of what I expect of you."

The Proposal Manager
"I've picked Joan Angelo because for a long time she's been threatening to turn into a project manager.[5] She understands our methodologies, our solutions, and she's one of our best computer center managers. *And* she knows databases. She can cost this job and staff it. She's a good listener. A good administrator. And they say techies hate people! That they can only manage computers! Beyond that, we've got to keep Joan happy so she'll continue as our softball pitcher. You think we're going to lose another game to Qwerty Systems?"

The Administrative and Production Team
"This is our King Kong contract, the one we're going to use to climb the Empire State building. Let's not slip up on some trifling detail.

"We want good I-dotters and T-crossers. From all fields. Good people from finance and accounting to help with the costing. Good contracts and legal people to formulate the final proposal as a part of the contract. Decent personnel people for recruiting.

"Here, especially, we also want good technical editors. People, I think God's malicious. He created thousands of brilliant engineers and programmers and deprived them of whatever gene they needed to communicate in English. Now, there are a few exceptions. But unless you're one, then you'd better let the technical editors do their thing. Help 'em get the information. Give them enough time to organize it. And make it persuasive.

---

[5] If Angelo isn't available, the marketing manager himself might also be a good candidate for a proposal manager—having already tracked the project and provided solutions. In many cases, the marketing man might even end up as the project manager.

"Above all," Kenner said, "let's use writers who can persuade. A proposal shouldn't just propose. Without sounding pushy, it should *sell*."

The Professional Staff

"We'll select them by experience and qualifications in the important areas. Like in managing computer centers and networks. We're talking about communications engineers. CPAs. Applications and data base programmers. Systems programmers. Computer operators. Maintenance people. And security analysts to meet federal Privacy Act regs."

The Red Team. "Our Red Team, as I'll call it, will try to discover weaknesses in our proposal before P&F does. They're the murder board. The ones who'll pretend they're with P&F. And they'll make sure about four things in particular:

1. That the proposal meets P&F's requirements.
2. That our technical solution is feasible and cost-effective.
3. That the technology methodology is right.
4. That the proposal meets our quality standards. Everything from format to language to completeness."

Concept #10: Take Care When Writing the Proposal

Here are the points that Kenner emphasized:

- Check and recheck to make sure you're responding to every requirement and spec in the solicitation. Figure out what's mandatory and what's just desired. Complete the response matrix.
- Make sure the proposal's contents are complete.
- Keep the writing understandable.
- Be certain you've understood the client's buying reasons and changed them into win strategies and themes within the proposal.

"Let's not lose sight of P&F's reasons—to get a system running quickly and reliably," Kenner explained. "And let's give proof that we'll come up with the best people and equipment for the job. That's our goal. Not to give P&F the best technology for technology's sake. Not to give them the cheapest system. Rather, to get them exactly what they asked for in the solicitation."

THE TYPICAL PROPOSAL

Cort's proposal included:

- A letter of transmittal.
- An executive summary.
- The response or compliance matrix—which would assure that Cort was addressing all the requirements and wishes expressed in the solicitation. (The sample appears on page 259 of Appendix B.)
- A table of contents. (See page 261 of Appendix C for a sample.)
- An introductory section, where you might introduce your company and lay out basic qualifications.
- An Assumptions and Constraints section. It identified the boundaries of the contract and assigned responsibilities and obligations of all parties—Cort's, P&F's, and those of suppliers and subcontractors. The section also cited risk potentials.
- An Understanding the Problem and Requirements section where Cort would lay out P&F's basic needs, problems and goals, and formal requirements. This section would show an understanding of them and tell how Cort's solutions would help.
- A technical plan.
- A work plan, describing the specific work steps that Cort would follow.
- A management plan.
- A listing of corporate and staff qualifications. Cort also would submit individual resumes of the top managers involved.
- Expression of management commitment.
- A document breaking down the prices for various parts of the job.
- Appendixes and miscellaneous material.

The Transmittal Letter

"I hate blah-blah transmittal letters," Kenner told his people. "I don't want the usual mush. I don't want us to write simply, 'Cort is pleased to submit this proposal.' And then my signature without anything else in between. No, ideally, we'll make this letter just as persuasive as the proposal itself. In fact, even more so. Can't underestimate these things as marketing tools. Yeah, we know that P&F's people associated with the project will read it. But it'll also go to others, higher up. Including some who'll never see the proposal itself. So let's not blow it.

"We'll begin with the normal boilerplate. The legal, technical, and contractual information that identifies the solicitation we're responding to. But then we'll go on to our most important ideas. The key benefits that P&F will enjoy.

"Let's tell how they can cheaply piggyback on our existing computer capacity. How we have regional centers all over the place that can help 'em out. We know that Qwerty doesn't. So without mentioning Qwerty, let's rub this in. And we won't just mention costs. We can also point out that we'll be up and running in three months. I suspect Qwerty will need a good six to nine.

"Of course we'll be brief. Just two paragraphs ought to do for telling P&F our special advantages and how they'll benefit.

"Later in the letter I'll include names and phone numbers at Cort. That is, who P&F can contact to clarify technical and cost matters in the proposal.

"And then at the end we'll identify enclosures or attachments. Including the main proposal itself."

See pages 200–201 for a Sample Transmittal Letter.

The Executive Summary

"Let's not kid ourselves. P&F's top executives and managers aren't going to be masochistic enough to slog through our whole proposal.

"So our executive summary must concisely explain what we're proposing. Any why our approach is superior. No gobbledygook, please. Remember, accountants, not engineers, reign supreme at P&F. We're not writing for hard-core techies. The people at the top of P&F don't know a bit from a byte.

Sample Proposal Transmittal Letter

March 12, 1987

James L. Lorain
Chief Purchasing Agent
P&F Enterprises
1000 Main Street
Tenley, MO 63552

Reference: Request for Proposal (RFP)#87–301, "Design and Install A Tax Preparation Network and Construction of Central and Regional Computer Facilities," dated January 5, 1987; and Amendment #1, dated January 28, 1987.

Dear Mr. Lorain:

The Cort Engineering Company is pleased to submit the attached proposal in response to the cited RFP #87–301. We have submitted the proposal in four volumes. Volume 1 includes the Technical Plan; Volume 2 includes our Cost Plan; Volume 3 includes the specifications for all equipment proposed, and Volume 4 includes Facility Plans.

Cort wishes to assure P&F that the tax preparation system will be installed and operational before the mandated date. We will use our current computer facilities and network; therefore, it will not be necessary to construct new facilities.

We have a staff of over 200 specialists in tax preparation systems and computer facility management. Moreover, we manage 15 similar facilities in the financial and tax area and have designed and installed tax preparation systems in 23 states on schedule and within budget.

Cort's offer should save you extensive time and resources since you will not have to buy additional computers or build facilities. We believe we can reduce your costs for tax preparation for each form by as much as 6 percent. In addition, we know we can help you achieve an optimal level of accuracy.

Our top managers have made a strong commitment to provide all the necessary corporate support to make this project successful. Even now we are busy preparing our facilities for you. To keep our commitment we are prepared to start work within two weeks after contract award, and to have the system available in three months.

If you have any questions of a technical nature please feel free to call Mr. Peter Warren, Vice President of Development. Questions on cost should be directed to Mr. Larry Dune at our company headquarters.

Sincerely,

Arthur Hill-Howe
President

Enc. Volume One (Technical)
Volume Two (Cost)
Volume Three (Equipment Specifications)
Volume Four (Facility Plans)

"Of course we'll have more room for elaboration here than in the transmittal letter. We can tell how we'll answer the more important requirements of the contract. And we can list our qualifications. For instance, our strength in tax preparation software. And our network of first-class computer centers.

"We can also get in a few subtle digs at Qwerty. Again, no name, please. We'll just say, 'We're ready to give you the system you need almost instantly. We won't have people scrambling to fit you in between other projects.' Which I suspect is the case at Qwerty, with that big Exxon contract they won the other day. Let's raise the overcommitment issue! Let's get P&F to evaluate our competitors as much as possible by *our* standards."

Item by item, Kenner listed what an executive summary should contain:

- Promised results and benefits to the client. Include the supporting evidence.

- Information about your strengths, the competitors' weaknesses, and why you're better.
- Mention of any special solutions to clients' problems, and why you consider such approaches to be the best.
- How you meet the evaluation criteria.
- A clear reflection of your win strategies that respond to the buying reasons.

The Response Matrix

Kenner didn't have to tell his people what a response matrix is. It lists, mainly in tabular form, the main solicitation requirements, which encompass at least: the statement of work, the specifications, and the evaluation criteria. Under each item you tell if you have met the requirement and where in the proposal the response is.

Some response matrixes—also known as compliance matrixes—might run longer than 50 pages and be a different document. After all, a good matrix will decompose the proposal sentence by sentence. You'll end up with a checklist to assure that you are indeed *responding* to equipment specs or other criteria. And you'll give the client an easy way of finding answers to his requirements.

See page 259 for an example of a response matrix.

The Introductory Section

Introduce your company. Explain qualifications. Sum up the relevant experience. Tell how committed your management is to the project. Show this. Say, "We'll start work within two weeks after the contract award."

"But," you may protest, "I've been dealing with that outfit for the last decade. Why should I introduce myself?"

Answer: because your client may have just hired an important staffer who, until now, has never heard of your company.

The Table of Contents

You should have at least four levels so that the contents can also serve as an index:

Level #1: Section titles ("Management Plan").

Level #2: Subsection titles ("Key Staff").

Level #3: Subsubsection titles ("Project Manager").

Level #4: Subsubsubsection titles ("Functions of Project Manager").

For brevity's sake, the sample table of contents on page 261 includes only two levels in most cases.

Assumptions and Constraints

"Don't underestimate the importance of this section, which helps keep the contract well managed," said Kenner. "Leave it out, and we could lose time and money. And so could P&F. After all, this section sets the boundaries of the contract and defines the responsibilities and obligations of the companies involved. It lays the foundation for costs. And for schedules. And for managing the contract."

"Let's say P&F doesn't meet their obligations. Then we'll have recourse to negotiate additional money, or maybe added equipment or more space. And so on. This is a fixed cost contract—for instance, fixed cost per so many hours of computer time. And this section helps keep things moving along the right way. It gives us recourse if people don't or can't live up to their responsibilities.

"It helps define conditions leading to possible price renegotiations. For instance, when fixed price conditions might be violated. Let's say P&F must decide what microcomputers they want delivered within 60 days after contract award. That'll help protect us against possible price increases by manufacturers. If P&F waits for 70 days to decide and prices go up, then they'll pay. We'll pay if prices go up before the 60-day period. Of course, we've explained why they probably won't.

"Write out a good Assumptions and Constraints section and in effect we'll be doing some terrific risk reduction research. And if the risks are too high—well, we'll restructure our proposal. Or maybe decide at the last minute not to go ahead."

With this and other possibilities in mind, Kenner's Assumptions and Constraints section ended up with two subsections, titled:

1. Client and Contractor Obligations, Including Constraints. Here he addressed such issues as what equipment and services the two companies would provide each other to get the job done. What deadlines must they meet? And what were the constraints—the limits on the scope and quantity of the services and equipment that Cort would supply?

2. The Constraints and Risks subsection—in effect, the "Snow in July" one. This would list unlikely situations or conditions that could wreak havoc on schedules, plans, quality management, and so forth. It would spell out the risk potentials.

You might organize your Assumptions and Constraints section differently. For instance, you may break the obligations part into (1) a subsection listing the equipment, materials, and services to be provided and (2) a subsection addressing issues such as deadlines.

Client and Contractor Obligations (Including Constraints)

Here Kenner's people outlined the respective obligations of Cort and P&F.

Under Obligations of Client—and this list is just the beginning—Kenner's people mentioned that at each location P&F must offer Cort:

- Parking, desk space, phones and secretarial service for Cort's staffers.
- Access both to the front door and loading docks. Never neglect the basics.
- On-site computer equipment, which Cort described in detail, just as he did some of the client's other responsibilities. Cort went beyond technical specs. It also told when it should have access to the equipment.
- Materials, such as computer paper.
- Several software engineers full time for 30 days to acquaint Cort's staffers with the existing P&F computer programs.
- Assurance that P&F wouldn't interfere with tests of equipment, and that if it did, Cort could pass the additional costs on.

- Prompt review of documents and automatic acceptance of them within a certain deadline if P&F did not find faults.

Cort emphasized that its whole bid assumed that P&F would meet those specified conditions and others. Then, under Obligations of Contractor, Cort promised among other things:

- Certain equipment, including some backup computers.
- The right kinds of people, some of whom Cort even identified by name.
- To meet deadlines for equipment delivery and services.
- On-site work for some tasks and remote work (via computer terminals) for others.

Under Constraints, Cort said P&F would give 48 hours warning of possible heavy loads on the system. Among other things, Cort also put a lid on the number of AT clones connected to any regional center. And it said it could not guarantee the quality of common carrier lines, such as those provided by phone companies. Otherwise, too, Cort limited the nature and scope of what P&F could expect.

The Constraints and Risks Subsection
The second subsection in Assumptions and Constraints would be The Constraints and Risks subsection: the Snow in July one.

More aptly, in Cort's case, it in effect could be the "Hurricane Section"; what if a hurricane blew down communications cables and otherwise interfered with work?

And how about other disasters—natural and man-made—that could raise expenses? Suppose traffic on the net increased by X number of tax forms transmitted for processing. Suppose AT&T made technical changes that raised costs. Or suppose a state legislature revised tax laws at the last minute and P&F could not change the computer system fast enough to use it to process returns.

Simply put, this section would list all the main risks and then lay out the consequential effects, who would be responsible, and suggested solutions. A more formal name might be Unlikely Conditions and Situations and Financial Consequences.

The Section on "Understanding the Problem and Requirements"

In telling his people how to write a winning proposal, Kenner had emphasized the importance of what you might informally call a "We Understand" section.

"Whether you're building a skyscraper or designing a computer system," he said, "you *must* understand a client's problems and needs. Otherwise you can't find an acceptable solution.

"And that's why we'll have a We Understand section. Otherwise called 'Understanding the Problem and Requirements.'

"Both informal and formal needs will drive our solutions. And we'll be more credible if we show how we arrived at our conclusions. And above all, show some sensitivity to our client's big concerns."

"We'll say we're aware of P&F's needs to get its network and processing operation going full steam as soon as possible. We'll marvel over the rapid growth of the P&F Tax Miser chain. We'll tell what could happen if P&F doesn't solve the problem. How growth could stop.

"Our proposal will explain how we identified the problem and arrived at our conclusion. We'll show our understanding of P&F's goals, its responsibilities, its fears, its limitations.

"And we'll relate our solutions to our findings here. While keeping in mind the client's buying reasons."

Formal Requirements
"Likewise, we'll give a good overview of how our solutions answer the formal requirements in the solicitation.

"We'll think about two things:

1. What our network and related processing services will mean functionally to P&F. They'll get their tax forms processed faster. And our system can split up the work much more efficiently than the present system.

2. What our solution will mean technically. The concept of what we're proposing. And how the system is configured. How the computers will hook up with each other."

The Technical Plan

Don't skimp on the technical part of the proposal. Granted, you rightly aimed your letter of transmittal and your executive summary at the nontechnical or semitechnical; top decision makers may be smart in their own specialties, but not in every one covered in your bid. But your technical section is different, for you're appealing to the hard-core specialists, the technicians, the ones who'll do the nitpicking for the generalists. Moreover, the technical plan provides guidelines for your own technical staff to execute the contract.

In more complicated language, your technical plan will carry out the same objectives that the other parts of the proposal did. It will try to turn the client's buying reasons into a preference for your solutions.

At Cort, Kenner broke the technical proposal into two parts, one dealing with design and the other with the execution of it.

Design
This subsection included:

- The configuration for the hardware, the software, the network, and the interfaces—both machine to machine and man to machine. The proposal described the industrial standards to which the design conformed.
- A breakdown of the design into the main system functions, such as communications, data entry, and fault diagnosis.
- A similar breakdown into applications functions like tax preparation, tax audit, and payroll.
- A description of performance and capabilities for each system function and each application function. The section covering the communications functions, for example, would describe the peaks and valleys of traffic loads between two points. For tax preparation data entry, the design would tell the number of tax forms that an average preparer would enter in an hour.
- Rationale for the design of the network and the related

computer system, based on such factors as how many tax returns P&F processed each day in different regions.

- The solution. Here, Cort showed how it would meet the performance requirements and network architectural needs by the number, makes and models of pieces of equipment, and in what configuration, and with what software. Cort would indicate the capabilities and capacities of each system component, matching this against P&F's requirements.
- Graphics showing the general concepts.

Kenner's people at this stage were offering the equivalent of a detailed sketch of a hot new car or a preliminary blueprint for a building.

In some cases the buyer might already have a design worked out, eliminating the need for this section.

If Cort had been designing a car, this is the section where the techies would have proposed the use of steel, aluminum, or plastic. They would have specified tensile strengths, too. Then they might have gone on to other details, such as the nuts and bolts of the production process. What kind of machinery, for instance, would work out best for stamping the car's sheet metal?

Implementation of the Design

This all-important technical methodology described *what* Cort's engineers would do and *how* they would do it.

For example, at the start, Cort would propose to perform a requirements analysis. (*Requirements analysis* is a standard services industry term that means the listing of all the requirements that a client says he needs to solve a problem. It can cover everything from electronic interfaces to the kinds of information that the system is to print out every day.)

Cort couldn't just say, "We will perform a requirements analysis and prepare a final report." Such a statement would be a technical *what*. And Cort would also have to address the *hows*—for instance, *how* it would do the analysis. That would include:

- What information Cort would collect for the analysis.
- How it would collect it.
- Why Cort needed such data.
- The sources of the information.
- What kinds of analytic methods it would apply to the collected information.
- How and when Cort would apply analytic methods.
- The reasons for using them.
- How to transform the results into requirements.

"If the solicitation involves a job where your task is to develop a design," Kenner had said before the drafting of the proposal, "then you present the technical methods and procedures you will apply to arrive at a design.

"Don't forget: In a solicitation where the contract will be to *develop* a design, you should *not* provide any design in the proposal. Many proposal writers mess up here. They mistakenly try to present a design before they actually perform the preceding requirements analysis.

"Take this case. A solicitation asks for a design of a training program. Then your technical plan must tell what you will do and how you will do it to create or design a training program. But many techies jump the gun. They develop a design for the training program without knowing what will be taught. That's a guaranteed way to rub an evaluation committee the wrong way."

Describing the implementation of the network design, Cort also included such items as:

- Descriptions of the methodology and techniques that the company will apply to develop and refine its solution. Among the methods or approaches could be the detailed analysis of user, functional and performance requirements; data requirements; program specs; software codes; and the testing of programs and communications connections. Cort would identify the language(s) used in any software it was to develop.
- An explanation of how Cort could use such special tools as prototyping techniques to speed up construction of the system, trim costs or offer other advantages that P&F in particular

would appreciate. Kenner's people always kept P&F's buying reasons in mind.
- Graphics, such as a map of the network and charts showing the flow of information.
- The general nature of the documentation to be provided.

Under certain circumstances, Cort might have submitted an entirely different technical plan that didn't go into the detail that this one did. What if the contract had been less extensive in scope or size? A 100-pound proposal with such a detailed technical plan would have been laughable for, say, the installation of a small local area network.[6]

Even when you go after mere $150,000 jobs, however, your technical proposal might thoroughly address the important issues related to each task discussed in the client's statement of work:

- The important goals. Cort might have quoted the statement directly, then told why its solution was responsive.
- How and what Cort needed to achieve the goals. Once more Cort would have cited language from the solicitation.
- Why Cort chose a specific approach, its benefits, and the expected outcome.
- Any potential risks in either the methodology or the results.

The Work Plan

In the work plan, Kenner's people gave a detailed, step-by-step breakdown—with schedules—of the work that Cort would do for P&F.

A typical work plan would identify each work step in the sequence needed to do each project task, such as requirements analysis.

A requirements analysis work plan could include steps like:

---

[6] Of course, Cort wouldn't have gone after such a tiny job in the first place.

Step #1: Collect all client operating procedure manuals.

Step #2: Identify and list all positions and tasks that use those procedures.

Step #3: List all information needed for each task for each P&F staffer.

Step #4: Review results of the initial analysis with the client section chief, who will want completeness and accuracy in the findings.

For the above steps, you would determine how much time you'd need for each task and list any resources required. And you would identify the criteria to verify successful completion of each. Augment the work plan material with a PERT-type schedule chart such as the one in Appendix D.[7]

Also add a work breakdown structure (WBS) to your work plan to list each piece of equipment or service that you'll need to meet the goals of your project for the contractor. You'll break down the work you'll have to do to deliver every product for the contract. A product could be a computer, software, a feasibility study, a procedure manual, an elevator, a building—in short, anything that your contract calls for you to provide the buyer.

Among other things, a WBS will help you troubleshoot if problems develop, by helping you trace their causes. A work breakdown structure also is useful for planning and monitoring the expenditure of resources, including travel and labor hours. Use block diagrams to show all equipment and material you need. And under each equipment-material category, you should identify all big tasks that you'll accomplish. For example:

1.1. Microcomputer Equipment. (Level 2)[8]

    1.1.1. Specifications Development. (Level 3)

    1.1.2. Acquisition. (Level 3)

    1.1.3. Installation & Test. (Level 3)

---

[7] PERT means "Project Evaluation Review Techniques." Among other things, PERT charts show relationships of time, sequences in which you must do steps in a process. For instance a PERT chart would make clear that you had to buy cable before you wired up your LAN.

[8] Level 1 would contain just the name of the project.

> 1.1.4. Connection of Workstations. (Level 3)
> 1.2. Computer Software (Level 2)
> 1.1.1. Specifications Development. (Level 3)
> 1.1.2. Compatibility test. (Level 3)
> 1.1.3. Acquisition. (Level 3)
> 1.1.4. Installation and Test. (Level 3)

Under each of the tasks you could offer another breakdown, a list of subtasks. In the case of "Specification Development" for the microcomputer equipment, you might say:

> 1.1.1. Specification Development. (Level 3)
> 1.1.1.1 Performance Requirements. (Level 4)
> 1.1.1.2. Hardware Functions. (Level 4)
> 1.1.1.3. Firmware Functions. (Level 4)

You should go down to at least five levels. At Level 5 you'd then develop a work package for each subtask.

The package might, for example, identify all direct and indirect costs related to people and their expenses.

This could include your staffers' salaries and their travel expenses—as well as listings for consultants and subcontractors. It might even cover the costs of the documentation needed to guide the people.

You could relate the parts of the work breakdown structure to each group of steps in the work plan. Then you'd gain a formidable project management tool. With it, for instance, you could:

- Remind your staffers what they must do on the first day of the job.
- Tell accounting, purchasing, and contracts what resources were needed to buy or lease, and what costs and when.
- Offer a reliable audit trail for contract auditing.

The Management Plan

Don't muddle the facts. A management plan isn't a technical plan or a work plan—its purposes and contents are different.

It will help answer the client's question: "Can you manage a $10 million contract?"

Simply put, a management plan will describe how your company will manage the contract, the design, or the study.

Its contents will include such items as:

1. Your corporate organization chart and the basic functions and responsibilities of various support organizations.
2. A project organization chart. It tells who'll manage the project, and which units will report to him. You also need a description of organizational functions.
3. A list of functions, responsibilities, and duties of key people on the contract. The client needs to know who to talk to to correct a problem or implement a last-minute change Also, this subsection can tell:
   a. Who manages your project budget.
   b. Who hires and fires.
   c. Who handles consultants and subcontractors.
4. A similar list of functions for lower-level people, key submanagers such as heads of engineering, training or quality assurance.

Your management plan should also lay out corporate management procedures. How will your company keep track of the project overall? Will it review the project in monthly documents, and perhaps monthly meetings? By what criteria? What about policing the budget and technical quality?

By laying out all those watchdog procedures in detail, Cort gave P&F a warm, fuzzy feeling, and convinced the client that top executives were carefully watching over the staff and subcontractors. Moreover, the management section also promised something of equal importance: regular progress reports for P&F itself.

The management section should include a staffing plan like the one discussed below.

The Staffing Plan

"This is our plan" Kenner had said at his proposal seminar, "for every job to be performed.

"It will include for every job or task:

- The kinds of people needed.
- The numbers assigned.
- Their qualifications, including years of experience.
- Their education.
- Justification for using them in those jobs."

Expression of Commitment by Top Corporate Management

"When you get married, you might give your wife a ring to show commitment," Kenner had said during the drafting of the contract. "We're not about to go to Tiffany's for P&F, but we might give them something meaningful in its own way—a formal expression of commitment.

"This section would include promises from top executives that their senior managers would give the contract full attention at all times to keep it on course. For example, a special audit team can be on the road at all times. And we can set up a special hot line to answer all questions. The section can promise we'll use our headquarters computer as a backup if a regional one goes down."

Corporate Qualifications

This section closely focused on the work at hand, and told of:

- Cort's financial history and growth in terms of revenue and staff size.
- General staff qualifications.
- Summaries of the work that Cort had done before on similar projects. Kenner's people rewrote the summaries to reflect P&F's special requirements.

Elsewhere Cort had already included the resumes or curriculum vitae (CVs) of key staffers on the project, rewritten to reflect the solicitation's key technical and management requirements.

Costs

In the price list for P&F Tax Miser, Cort included costs for use of computer resources, reflecting:

- CPU[9] time.
- The length of usage of the input and output channels of the system—that is, how long P&F took to send information to the regional computers and how long it took to receive the data.
- Disk storage space and time usage.

Under other circumstances, bidders might have justified their offers by breaking them down by years into these categories:

- Labor. This item would include overhead, general and administrative costs (G&A), and profit. The others would cover just G&A and profit.
- Equipment rental, leasing costs, or purchases.
- Materials.
- Travel.
- Documentation.
- Subcontractors' costs.
- Consultants' costs.

The Appendix

This section included backup material, for instance:

- Samples of work done for other clients, such as studies, designs, models used for design, detailed technical specs.
- Reference materials too bulky to include in the main body of the proposal.
- All specifications for equipment to be delivered, including original manufacturers' literature and a bill of materials for all equipment (to simplify, a bill of materials is a very detailed shopping list).
- Executed copies of subcontracts and consultant agreements.

---

[9] The central processing unit is a major part of a computer that manipulates information.

SUMMING UP

Cort realized that proposals are marketing tools—not just technical documents to slap together without any appreciation of their potential powers of persuasion. A good proposal won't automatically win business, but it's a big boost. Ideally it will be a fitting finale, the culmination of the zero defect marketing that began long ago when you first heard about the lead. *Proposals are the only means to win future business.*

Of course, having won services contracts, you might not stop there. In satisfying your customer's needs you may decide that you've invented a product that you wish to offer elsewhere. If so, however, you'll do well to avoid the Mud Pie Syndrome described in the next few pages.

Do's

1. Plan, plan, plan, and replan your proposal. Write a proposal plan as if you were writing equipment specs. Let everyone know what to write.

2. Read and reread the solicitation until it's completely understood. Then read it again the day before you submit the proposal.

3. Identify your proposal's capture and win strategies at the very start—that is, when you first talk to the client and grasp his requirements and buying reasons.

4. Write a proposal for people of all technical levels, and for the nontechnical person who'll pass on it. Alas, you must write, too, for different levels of business sophistication.

5. Develop a proposal production style guide to tell writers how to structure and format their material. You can also give stylistic guidance and standardize technical terms.

Don'ts

1. Don't use a poorly written proposal as a model just because, by some fluke, it won a job.

2. Don't make the proposal a technical tutorial unless the client requests it. Stick to essential explanations.

3. Avoid risky or novel solutions without strong evidence of workability, reliability and cost-effectiveness.

4. Don't use unsubstantiated claims—technical or otherwise. Don't say, "We're the largest company in the field," if you can't point to a statistic from a credible source.

5. Don't look at proposals as mere sales literature. Yes, they must sell; that's the reason for the chapter title, "The 100-Pound Sales Brochure." But the proposals also must present a major, workable plan. They aren't just four-page sales fliers.

CHAPTER 11

PRODUCTS: MUD PIES
FOR SALE

Those Philistines were still struggling.

If contracts kept eluding them, they'd indeed have to retreat from that glassy edifice overlooking the Potomac. Morosely the Philistines recalled the fate of a stricken rival, Belsk Systems: dead at age 20 in a windowless warehouse in a seedier section of Prince George's County.

Seymour Turner, Philistine's marketing V.P., hated the thought of forsaking Thai restaurants at lunchtime for second-rate steak houses. Suddenly, however, an idea dawned: The competition for services contracts was heating up, so why not jack up the cash flow with a hot, high-profit product? Why not sell the new software that Philistine had designed for the Delcom Corporation? Delcom had loved it. The Philistines had come up with a series of library-style programs to keep track of other programs and also help debug them. So why not get into *products,* not just services? Surely other firms would line up to buy the Philistines' brainchildren. And if the library program worked, why not the hot new computer keyboard that Philistine also had designed for an appreciative Delcom? Instantly Philistine might become a product-oriented company. No longer might Turner and colleagues fret over the ups and downs of services contracting.

Turner's first mistake was to quickly hire the Silver Fox, a top salesman from Comprod, a multimillion-dollar software publisher.

The Fox named the library program "Marian." He bought a

mailing list with the names of most of the mainframe computer users in the eastern United States. Then he and assistants knocked on doors, leaving evaluation copies behind. In 14 months, however, the foxes licensed only three copies; $300,000 poorer for the experience, Philistine fired the whole pack.

This fiasco, however, was hardly the foxes' fault. Philistine had been selling mud pies. Remember them? Your sister might have made mud pies for a loyal client, your mother, who gladly paid a penny each. This customized product, however, wouldn't have fared so well at the local A&P. What works for one buyer won't wow them all. Marian, for instance, excelled at finding bugs only in certain kinds of programs. What's more, rivals had been selling two identical and superior products for more than five years and had captured over 97 percent of the market. No niches existed.

Alas, many otherwise smart firms in high tech services are making mud pies. They think hundreds or even thousands of companies will appreciate the customized solutions that a few clients loved.

How not to perpetrate a mud pie? Here I'll discuss:

1. *Why service companies don't develop and market products well.* I'll describe common kinds of mistakes. Some companies overestimate the appeal of customized solutions with which they're familiar. They think that a product demonstrating technical feasibility will also offer the magic of commercial feasibility. Other services companies don't even build products they understand. They just read of a demand for hardware, software, whatever, and rush in—without enough capitalization or appreciation of market needs and quirks.

2. *A philosophy of innovation.* The right philosophy can help breed innovation. Committees don't invent. People do. Adopt this philosophy and make your corporate bureaucrats respect both it and your innovators.

3. *The basic kinds of needs to which inventions respond.* If you understand the relevant psychology, you might just grow richer off technology.

4. *A success story at the Wicket Company, an engineering firm.* It invested in and successfully marketed a laser-style device to detect chemical leaks.

5. *Seventeen steps to take you from a product idea to the marketplace.* You'll learn in detail how your own company can invent and market a *good* product.

6. *Some final thoughts—on global Snark games.* Whether you're marketing products or ideas, you're part of international competition. The futures of your company and your country are more intertwined than you might think. Be a patriot! Invent!

Without proper development, however, even the most revolutionary inventions can lose money. And here you'll learn how to write a proposal to guide your own company and show investors that you know what you're doing. Also, you'll find out how to build a "success model." It will help point you in the right direction—toward a potentially lucrative market.

Don't misunderstand me: Products *can* make money even for a few services companies—sometimes $5 or $10 million in gross revenue in a few years. You won't triumph in the products business, however, if your wares simply give you a steady cash flow to balance out the vagaries of the services market. There are too many risks. Instead you must breed a whole herd of cash cows that you can milk for more than half your income. Your goal shouldn't be one money maker but rather the creation of a family of related products for the mass market.[1] With life spans of maybe a year or two, high tech products quickly are ready for the junk pile; just look at how MS-DOS computers killed off most new CP/M machines for the office.

Still, for companies trying to come up with their first new products, the original investment in research is sometimes next to nothing—at least in the case of software. You *may* have already done the basic R&D in your quest for customized solutions. The technical expertise so often is already there.

The marketing skills, however, may not be present—especially those that can help your lab whizzes invent winners for the marketplace. And often the marketing expenditures may end up dwarfing the R&D-related ones.

---

[1] The products would be "family" in the sense of working well together—for instance, a database, a word processor and a spreadsheet that could all read each other's files.

THE "BEWITCHED AND BEWILDERED COMPANY" AND OTHER RED INKERS

Philistine was a "Bewitched and Bewildered Company," one of five categories of firms with losing products. Others are "The Cross-Eyed Optimist," "The House of Straw," "The Bumper Cropper," and "The Alchemist."

The Bewitched and Bewildered Company

A Bewitched and Bewildered company can just taste all the money that people in high tech are raking in. "Why, just look how Lotus made out like bandits with 1–2–3," a typical B&B executive will cry out. "Boys, this is one bandwagon we don't want to miss." Or a B&B will pester his techies, asking project managers if they haven't developed something fit for mass production. Consider what happened at the Bludsoe Corporation, a cash-hungry company that hoped to beef up its profits enough to make it an attractive acquisition prospect.

This professional services company developed business information systems, accounting systems, and some personnel management systems. Executives there in the mid-'80s thought that the time was ripe for a move into the products market. Bludsoe's president formed a Product Search Committee (an animal sometimes known as the Product Development Committee or the Product Review Committee). He appointed as chairman a manager named Jason Taylor, whose only experience with products was with those he bought at Safeway or Toys "R" Us, Inc. That seemed inevitable. Bludsoe hadn't ever sold products before, just services.

Taylor's Product Search Committee didn't prepare any plan for selecting marketable products. It worked by the rule of "ad hocery."

The committee rejected dozens and dozens of new product ideas, and it also turned down a plan to buy a company that had a product. But Taylor's people did decide on an alternative. They would start a research and development center, which a Ph.D. with a research background would manage. The Ph.D. was to develop new products that Bludsoe could market. There was a constraint, however. The R&D center must be self-sufficient, self-supporting. Clients must fund it on retainer or the market place must provide the funds; that is, it would be a contract research and development operation.

After spending three futile months searching for the right person, Bludsoe hired a Ph.D. in biochemistry, Max Ferguson, a recent retiree from a commercial laboratory, and a member of Taylor's church congregation.

"This is a first-class opportunity," said Taylor. "You can do what you want. Why, you can even develop projects in your field."

"What am I going to do," Ferguson joked to himself, "develop an organic computer?" But he was bored enough with retirement to sign up. Besides, Taylor didn't care what kind of product Ferguson created—just so it would fill up Bludsoe's coffers.

Max went looking for R&D contracts, only to hear his old friends laugh at him.

"Look, Max," they said, "Bludsoe isn't a chemical company. What can you do for us?" And Bludsoe's own clients were a blank to Max. What's more, they had no understanding of, or even need for, research and development. On top of everything else, Max didn't even know what to discuss with them to be able to prompt some requests for research and development. Eight months passed. Max didn't have a contract in sight, so he gave up—hating himself for having taken the job.

To this day, Taylor still wonders why Bludsoe couldn't crack Products.

The Cross-Eyed Optimist

Cross-Eyed Optimists believe that their products will sweep the market and make them millions and millions for years—starting next week.

They're often as blind to the inevitable march of high tech as they are to their babies' flaws. Alas, the Optimists are all too typical of professional engineering services companies that try to crack the competitive products market.

Normally, a Cross-Eyed Optimist (a company or a person)

starts with a product created for one client. That happened at Starke Engineering, a $10-million-a-year shop that designed customized switches[2] to help computer systems route messages to the right places.

The Intercom Co. had itched for the switches. It had issued a contract to Starke for Starke to design them and have 30 made. Starke had kept the commercial rights. "So," thought some of its executives, "why not sell to other companies? Why shouldn't Starke itself mass-produce products?"

Don Wesley, however, the young pup who was Starke's marketing V.P., didn't grasp the difference between marketing a product and marketing services—an error also committed by his seniors.

They couldn't appreciate the credibility that already-established competitors enjoy even in a fast-moving market. Wasn't Starke's product new and different; wouldn't the world suddenly love it?

So an excited Wesley met with his bosses to decide what to do with the Product.

Starke's president asked for a pro forma statement[3] and even a calculation of return on investment. That was about it. No one even understood what was going on, let alone how to spell "pro forma," but they tried. Even the finance vice president couldn't help. He had only developed documentation and financial statements for the services market. The V.P. hadn't any idea, for example, of what to include in a pro forma statement. No one appreciated the ingredients of a business marketing plan for a product going to market. Even the V.P. didn't know how to calculate product cost and price.

Still, the people at Starke ran a spreadsheet off their local computer—even if they could barely tell whether an ROI[4] of 3 percent was better than one of 300 percent. Anything over two digits scared them. They worried that the prices would be too high for the product to sell or too low for them to turn a profit.

Their confidence, however, never waned. Most of all they

---

[2] Made by subcontractors.

[3] A pro forma statement is a financial spreadsheet that lists by time period the expected sales, the number of units sold, the operating expenses to make all the units, sales expenses, profit margins, R&D costs, general and administrative costs, and net profits.

[4] Return on investment.

believed in Product Integrity. Except for a new name and the
addition of a blue and gold wrapper, they didn't change the original switch. Even the documentation remained the same. Hadn't
Intercom loved everything?

The next step toward the forthcoming debacle was to hire a
salesman. Pity him. Bereft of good marketing support, he had to
puzzle out not only who the clients were, but also how to price the
thing. He didn't even know when the first switch would roll off
the product line. Starke hadn't even picked a manufacturer.

Confused, the salesman called the product council together and
asked their opinions on technical specs, pricing, and delivery
schedules.

Young Wesley and colleagues toted up what their client's costs
had been for this one-time, one-shot product.

They computed engineering, material, and assembly costs,
added their profit, and, lo and behold, came up with a price of
$250,000 a switch. They completely ignored both market demand
and production economies of scale.

Not surprisingly, prospective customers balked. They asked
nasty questions.

"Have you done any general risk assessment?" inquired an
astute man at Ford. "Have you compared the costs of having your
switch versus the cost of not having it in our plant? So what if we
lose out on some information or it gets routed to the wrong places?
Are you sure the loss will be $150,000? Sell it for $60,000 a unit
and I'll consider, except that I'm a mass buyer, so you'd really
better make that $50,000."

Wesley was speechless.

"And how about a warranty for your new miracle?" the Ford
man went on. "And don't you have any reliability records to back
up your claims?"

A dispirited Wesley consulted with his engineering, production, and administrative people. All around Starke, faces
blanched. No one was really familiar with this kind of product
development, marketing, and manufacturing.

This charade went on for six months. Meanwhile, irrationally,
Wesley looked forward to jumping into other product lines; wasn't
it the divine right of every marketing V.P. to have his own product line? To the end, he denied he'd belly flopped.

There are hundreds of men like Wesley. Oh, their companies
may get their product developed, they may get somebody even to
build a prototype or manufacture part of it, but you can be sure
that they're not going to be too successful unless they change.

First step might be to fire Wesley if he doesn't wise up. Otherwise, a year or two later, when he wants to trade up to a bigger BMW, he'll again try to make a name for himself as a products man.

The House of Straw

Ralph Cassworth, president of a House of Straw-style company, hated to invest or plan and paid dearly for this disliking.

Cassworth loved "economy"—his word for almost malevolent cheapness. Cassworth pestered staffers to spend their spare hours developing new products or modifying the ones that they'd customized for clients. And that was true even when billable work alone was overwhelming the techies at Cassworth, Inc.

Cassworth gave "Management by Wandering Around" a bad name. Say, you were an auditor and he wandered into your office. Before you knew it, he'd be barking: "Why aren't you at work designing that new manufacturing process for the product we discussed?" Or suppose you were a corporate contracts director for services; then Cassworth would demand that you work until midnight to draft legal documents or dealer agreements. People's jobs of the day seemed to depend on where Cassworth took his walk.

This is the same company that indulged in such pleasantries as:

- Using marketing staffers to push furniture around if they were moving from one part of the building to another.
- Ordering $60,000-a-year engineers to do administrative work because they were on overhead, rather than assigning them to research and development. At times they even had to edit underlings' resumes. Instead, Cassworth should have hired editors for such tasks.
- Trying to mooch free parts of equipment off suppliers. A few complied. The rest wanted money. The resultant delays prolonged development of prototypes.

Simply put, the people at Cassworth lacked both time and the sense of direction they needed to develop successful new products.

The Bumper Cropper

The Bumper Cropper invests in a product that is already part of a bumper crop. The Bumper's offering will compete against several products that have dominated the market for years. Already, the storage bins overflow. And Bumper isn't running a wheat farm with Uncle Sam providing supports. So the Bumper is out of luck when customers decide that the product is "me-too-ish." Even the niches may be full. The technology may be mature. And if it isn't? Customers may still favor familiar names, and old-timers may enjoy better distribution and yet other advantages. Without a significant price or performance advantage—or the ability to fill a niche well—a newcomer can't buck the odds.

Sil-Com Services, a custom designer of computer systems, lost more than a million dollars because it failed to grasp those simple truths.

For years, Sil-Com Services had ignored the market for IBM mainframe database management systems. One day, however, a trade magazine ran an enticing headline, "BOOM MARKET FOR MAINFRAME RELATIONAL DBMSs" and suddenly Rex Hamilton, Sil-Com's president, fixated on the idea of redeveloping his company's DBMS for the mainframe market. Hadn't Sil-Com designed and delivered a similar DBMS for a client several years ago?

And so Sil-Com came up with a supposedly "improved" product of its own. The company added a few query and security wrinkles, but otherwise the DBMS was a clone of everything else on the market. Unfortunately, rival manufacturers had been selling their own DBMSs for more than a decade and had cornered over 97 percent of the market. What's more, they had better reputations and the same price, and prospective buyers asked such nasty questions as:

"Who are you, and when did you get into this business?"

"What other products have you sold and serviced and supported?

"What is your record?"

"Why should I switch from XYZ DBMS? We've had 'em around for three years, and their system works well. And all our people are trained on XYZs, and wouldn't want to change. Why, there's even a local users group. In fact, I'm vice president. Plus XYZ offer superb software maintenance. Their product is first-rate. It's good now, and they're making it better. They got the bugs out long ago. Your DBMS is untried. What else do you want me to say?"

The Alchemist

For much of Western history, some dreamers had tried to turn lead into gold.

Today we have corporate dreamers—Alchemist-style companies that think they can take a mundane material or product and make customers both need and crave it. They typically create products in search of a problem.

Invent a windshield wiper for eyeglasses and the "Alchemist" description will fit you. It happened. "Alchemist" would also describe the small manufacturer of what supposedly was a highly absorbent, low price cloth. The firm hoped to turn the product into a winner—by making it waterproof and floatable as a gimmick to attract consumer attention. There was a problem, however. If the material had to be absorbent, how could it also be waterproof and floatable? Shortly after the advertising campaign the company went out of business.

In both those cases, the innovators depended not on reason but on their intuition, which is fine if you're Alexander Graham Bell or Thomas Edison but not so commendable if your mind and inventions are more prosaic. Ordinary mortals must plan; they must assess the marketplace and its requirements.

A PHILOSOPHY OF INNOVATION

Imagine a modern committee guiding Thomas Edison. Advice:

Committee person #1: "Nice idea, Tom, but our gas lights division is worried that your work will hurt sales."

Committee person #2: "My brother-in-law, who should know, says you'll never be able to keep a carbon filament going for more than an hour."

Committee person #3: "Mr. Edison, we haven't seen a single penny of profit from this foolishness. Our stock's down and you have the nerve to ask for *more* time? Just what will we tell the analysts?"

Committee person #4: "I don't see the point of inventing an electric light bulb. Why, the Japanese will have a knockoff out within a year."

I've invented products[5] and have read millions of words on the history of technology, and one truth emerges supreme: *Committees do not create good products.*
Why? Because:

1. They are not the buyers, or the consumer, or the marketplace.
2. Committee members rarely understand how innovators can perceive people's needs and what solutions can satisfy them at the right time and the right place.

There is an art involved in being an innovator. Normally a product is the brainchild of one person with the right experience, the right knowledge, the right insight, to create something that did not exist before. Later the product will show the essential influence of others, including customers who try out test versions. But the original creation normally comes from one man or woman.

In this chapter, I can't give you a foolproof system of designing a marketable product. But I can help you think more creatively—and nurture those to whom this comes automatically.

You can encourage innovators to stay on schedule, for instance. Often they want to jump steps, go backwards, continually refine, rebuild and redo their product; and without exercising heavy-handed controls, you can help them focus their efforts better. That isn't easy. The line between frivolity and directed curiosity can be thinner than you'd care to imagine. In fact, paradoxically, even R&D "failures" can offer benefits. They help eliminate possibilities on which, without research, you might waste resources. So show understanding toward innovators.

Above all, you can help them and their managers to create products that will both meet customers' needs and turn a profit. Bear in mind that overnight successes rarely happen.

It takes time, energy, and resources to bring something to a successful stage—six months, a year, two years. If you are the first with this new product, it will be expensive and an uphill climb, no doubt, and the more successful you are, the more imi-

---

[5] Everything from educational board games to microfiche readers to systems for warning chemical plants of toxic leaks.

tators you will have. Your product's life as it stands at the beginning will be no more than one year, and maybe even less.

If your company lacks innovators and a first-of-its-kind product, you might still succeed—if by the end of that year you can find the right niche for an improved version of an original. You must refine the original product, however, so that:

- Your version isn't an exact duplication of the original.
- Your price is lower.
- You meet other ground rules of marketing and marketing management—from sales promotion to distribution, from marketing management to contract management, to purchasing.

This whole game is an art. It takes a virtuoso to develop the product, and a company with the same qualities to help it bear fruit.

THE BASIC KINDS OF NEEDS TO WHICH INVENTIONS RESPOND

Some professors of business and marketing do not believe that the need comes first. In fact, they believe you can create the need. Of course, if you believe this, then you will operate like the Alchemist.

Alas, the experts are blind to the bridge between the product and the buyer—the intuition of the inventors and innovators themselves.

Those with the sharpest intuition are the stars, the equals of Edison, the Wright Brothers, Bill Lear (he invented the LearJet), and Steve Jobs. The best stars can understand needs and solutions ahead of time. Only after you've developed and sold the product do you know what particular need it satisfied. The dream product's sales will reach the hundreds of millions. And that will be just the start. Ideally there'll be potential for improvements—and more sales. This will be a market, not just a niche.

Most companies, however, lack a resident Edison or Bell. Before attempting to conquer a market, they should try filling a niche. Even this, however, requires a good understanding of the needs that drive buyers. All products reduce stress in some way. They help you live with stress-producing conditions—for example, lack of knowledge or shortages of time and money.

But you can't just say, "My product will reduce stress." Rather you must examine that need as it exists at three interrelated levels:

- The primary level. It's the need for sustenance and survival.
- The second, or psychic, level. That would include the needs to feel secure, in control, and able to impress others. This area encompasses some other social aspects, such as being able to coordinate, cooperate, and protect, as well as consequences, or cost of consequences, of using an existing technique or product that will drive people to look for an improvement or an alternative product.
- The third, more intellectual level; here we're talking about goals such as cost efficiency, or performance efficiency, or utility, or timeliness, or precision and accuracy.

To discover such needs, you should first look for signs of unsatisfied needs: indications of gaps, problems, or constraints that exist despite people's contentions that they're happy with the status quo. You can learn of the existence of the gaps in a number of ways—ranging from articles in trade newspapers to speeches at conferences.

You'll next have to come up with a list of requirements that your solution must satisfy; then you'll have to convert those requirements into a product.

Meeting the Needs

You can gain acceptance by exploiting one of these three driving forces: (1) universally acknowledged demand, (2) latent needs, or (3) institutionalized needs.

1. Acknowledged demand. Here, there's a consolidated demand to produce a new product to meet needs, so you either invent a new product or improve an existing one.

No matter what the price of gas, we'll always think it too

high. Who doesn't recognize the need for a good 100-miles-per-gallon car?

2. Latent needs.   In this case, an innovator or inventor understands and then develops a product that excites the dynamics of the needs. A great demand follows.

Take the Sony video cassette recorder. It offered direct access to entertainment more efficiently and cheaply than movie theaters did, thus reducing various kinds of stress from shortages of time and money.

In some cases, however, you must create a bridge between a product and the satisfaction of a need. No one at first knew what to do with lasers. Only later did we learn how valuable they could be as a tool for surveyors or surgeons—applications that continue to grow.

3. Institutionalized needs.   Here you're seeking more efficient and effective ways in which people can do their jobs—so you look at:

- How they make decisions and interact.
- The weaknesses of the present methods.
- The consequences of these weaknesses, in terms of cost, time, and competition, among other things.

This "needs" approach can offer a treasure trove of new ideas. But you may run smack into formidable roadblocks—the biggest of which will be people's reluctance to replace existing practices and processes.

Remember: The prospective client or company has organized its people and operations around current practices. Your customer has invested time, money, even emotion, in the status quo.

HOW WICKET CORPORATION CREATED A PRODUCT

You can improvise a marketing campaign even if your company lacks the usual resources and people. What's more, you can obtain feedback from clients early enough to influence the design of the final product or at least its immediate successors.

An example is the Wicket Corporation, an engineering services firm.

Wicket had a yearly income of $60 million, a small part of which came from a division bought several years before. It specialized in laser-type devices used in communications, engineering and construction. One of the stars there was a bright young engineer and physicist named Andy Anderson. And bells gonged in Anderson's head when he read of the ability of lasers to read certain suspended particles and identify their composition.

Why not put a laser device together with a microprocessor that could identify accidental or fugitive atmospheric releases of toxic chemicals? What a natural for chemical plants!

And so Anderson spent hundreds of hours writing some prototype software that could identify what particles acted on the laser beam. He also developed 3D graphic displays showing the speed, direction and altitude of the leaking chemicals. In emergencies, chemical plants could use this information to direct the evacuation of people in the path of the toxic clouds.

Two months later he had a working model, which he field tested. He named it "Sherlock." And he showed it off to Wicket's president, chief scientists, and engineers.

"OK, so you've got it working," said Fred Wicket, "but now we've got to make it marketable."

Wicket, however, hadn't budgeted funds to do a comprehensive market survey. In fact, it even lacked a product manager or a sales staff. Until now, Wicket had developed its laser products on direct order from the clients. Wicket wasn't exactly Procter and Gamble when it came to marketing.

But Bob Beaumont, the company's marketing director, was quick with a suggestion.

"Let's have the marketplace develop the survey for us," he said. "At the same time, we can test the marketplace's interest and demand. Hell, maybe the chemical manufacturers already have something like this. We'll find out soon enough whether we have something marketable. Why, the marketplace will even tell us how they would like the device configured, and what price they want to pay."

"But," protested Fred Wicket, "how in the world are we going to keep our costs down?"

"Two ways," said Beaumont. "First, I'll visit the Chemical Manufacturers' Association and ask them what they think about the device. I'll even take them out in the field and demonstrate it for them.

"Second," he went on, "I know Richard Shortree, the chemical industry's most trusted environmental engineer. Now, he happens also to be manager of a medium-sized chlorine plant. Let's see what he has to say." So Beaumont and Anderson visited both Shortree's plant and a chemical engineers group.

"Just what I'm looking for," said Shortree.

"You'll save people a bundle," said an association representative. Chemical plants were tired of paying too much for inadequate chemical detectors and for lawsuits from workers and from people living near by.

With hardly any prodding, Shortree and the association staff offered suggestions about how to package the Sherlock system, how to configure it, and how to present the product. What's more, they offered a ballpark price that they thought the industry would pay for the product. They also suggested that Beaumont demonstrate Sherlock at two upcoming conferences which, in fact, Shortree helped to arrange as he was on the board of both associations.

The response to the demonstration stunned Beaumont, Anderson and Wicket.

In two days at the first conference, they collected 550 tentative orders for Sherlock—along with comments by the potential buyers as to (1) what chemicals they worried about, (2) what chemicals should be covered in the system's software, and (3) what prices they would pay. In fact, most of them said that they would really prefer to lease Sherlock, rather than buy it outright. That way, they could have the system without going through corporate headquarters to get an OK on its budget. "Something else," a plant manager said. "How about integrating Sherlock with our current safety systems?" Wicket nodded. The plant manager's idea might lead to more than a million dollars in extra sales.

The second conference two weeks later yielded similar results with almost 275 industrial respondents.

In fact, Wicket collected feedback and interest from a total of 825 potential clients altogether, a sample of almost 10 percent of the total population of chemical plants in the United States.

A week after Wicket and the others returned from the second conference, they received 11 purchase orders for Sherlock. They were only in a prototype phase. And yet they'd collected and compiled a comprehensive and substantial market survey. At the same time they had carried on a cheap, effective public relations campaign.

Wicket had actually done better than if they had conducted a standard market survey, which wouldn't have presented the product in its real form, in its operational form, to such a large audience. What's more, Wicket wouldn't have enjoyed the kind of feedback from potential buyers who put their hands on the product and were able to picture how they could use it.

Simply put, there was indeed a hidden need for Sherlock. Plant managers did have a brute-force, manual detection system that apparently had been working for them. It was slow, however. And it was not as thorough and reliable. But it was all that the plant managers and other potential customers knew. No demand existed for Sherlock. But a *need* did. Otherwise, the chemical giants wouldn't have been itching to buy the product.

Wicket, of course, is just one example of how a services company can develop and market an invention. There is no official system. But there are ways to reduce the possibility of failure.

FROM PRODUCT IDEA TO MARKETPLACE: SEVENTEEN STEPS

How to emulate Wicket rather than Philistine? Suppose that a hot, young scientist like Anderson hasn't approached you with a product idea. You're starting from square one. How to make the transition into products?

Step by step, here is how you *might* conquer the marketplace eventually or at least slip into the right niche. Take nothing for granted except the possibility of failure for the unwary.

With prudence, however, you can keep your risks down.

Step #1: Hire Creative People

A company's ideas are only as good as its people. Forget what the psychologists claim—there's no one test for creativity. What you can do, however, is to hire people with proven track records. They've written seminal articles or books or, even better, have themselves invented or developed new products. Significantly, creative people needn't be attuned to specific market segments or niches. A first-class designer of microcomputers, for instance, could thrive at a company making local area networks.

Although creativity should be one of the criteria for hiring, it should be far from the only one. Some of the least creative people, in fact, are the best doers. In certain routine jobs, creativity may even be a negative.

At the professional level, however, you don't want drones and "yes" people. Ideally, everyone should show at least the potential for fresh thought.

Don't give up on people you have already hired, of course. Think of them as uninventive bores and they'll act the part.

Step #2: Get Out the News that You Want Ideas

First find out what products your staff has in mind. As their boss, you must be careful not to crimp their imaginations at this beginning stage. Your people may think of possibilities that would never cross your mind. Then, shortly, you can help them focus their imaginations in the right direction, but first let them trot out their own brainchildren—as long as they're in areas related to your company's strengths.

Respect serendipity. Think of Dr. Fleming and the spilled mold that led to the discovery of penicillin. At the same time, realize that most practical inventions start with needs. Ask in memos and at special meetings, "What problems need fixing? And what solution would mean a big improvement? What bears investigating? Anything related to the services we offer?"

Suppose your firm custom designs precision instruments and you want to break into the consumer market with a product for the growing number of elderly people. Then your people might think about the possibility of an electronic pill dispenser for people who must take many kinds of pills daily. This device could disgorge the right amounts of medication at the right times.

Or perhaps your company wants to solve the problem of building cheaper, better houses. Then how about electronic measuring tools that would be faster and more accurate than rulers and maybe even plumb bobs? That might save half of certain labor costs in the construction of houses.

Maybe, out of a hundred ideas, one or two might show promise, but, without an open mind, how can you be sure?

Step #3: Create the Check Version of Your Product, and Describe Its Capabilities

A check version—a real, live, working model—doesn't include all the bells and whistles of a final product. Just check key features. Verify, in the most meaningful ways, that your creation will work as well as you expect.

As you create the check version, you should simultaneously:

- Spell out all its features and capabilities.
- Tell how it will be used.
- List the benefits it ideally will provide.
- Indicate what it will replace, and what conditions it will improve.

You don't need specifications at this time for hardware, materials, or software. If you do have a hard product, you might offer drawings or show a flow diagram of the technological process you have in mind.

Step #4: Pick Outside Brains and Do Library Work

Got an idea yourself or from an employee? Then talk to *trustworthy* leaders in your own industry who are not competitors and also to leaders in the prospective client community.

"Leader" doesn't mean your best friend or an obscure manager. Nor does it mean venture capitalists. That's for later.

Rather, leaders make the rules and guide the industry. They themselves may have succeeded either in bringing products to the marketplace, or in backing new product development.

When you talk to these leaders, ask them to sign a nondisclosure agreement to protect your idea.[6] Ideally take a few to lunch or dinner. If not, how about at least a phone conversation at a mutually convenient time? Offer a consulting fee if you need more than 10 minutes.

Ask:

- Is the idea good?
- Is the time ripe for development and marketing?
- Are there any competitors?
- Does this product have any weaknesses?
- Who will be the best clients, by industry, or by niche, or by segment?
- What do you think a reasonable price would be for people to pay?
- How long do you think that this product will last before it will be superseded by someone else's product or improvement?
- Do I really have the right application, or is the product too clumsy? Will there be usage problems, or maintenance problems?

Also, go to professional association conferences to feel out reaction to your idea. You *do* have contacts of this kind, don't you? You should. You want them anyway, as prospective clients for your firm's professional services. You want to talk to them to find out what their needs are, both product- and service-related.

Similarly, you should pore through technical literature to see (1) if competing products exist and (2) how you can improve your own brainchild. Seek out articles on techniques that might have some bearing on the product in question. For example, an article may discuss a new chip being developed at Bell Labs that might eventually go into a competing product, or might even be available for you to incorporate in your own product.

As vividly as you can, you're trying to profile the technical and commercial feasibility for the new product and to answer

---

[6] I'll say "your" for the sake of brevity. The actual idea may have come from an employee.

basic questions about it. How should it look, feel, and operate? What will your own manufacturing costs be? How much should you charge customers? And how can you refine the product, once the first version is sliding off the assembly line?

Also, check your local reference library or information service to get an idea of the universe of possible buyers, such as industries and schools. How many plants are there in the industry to which you hope to sell?

Step #5: Start Thinking about Details

Plug in the feedback you've gotten from industry leaders and from your library research.

What will your product's general capabilities and features be? What materials will you use? What will it look like?

You aren't yet firming up specs. As input for this paper model or for your check version, you are using the feedback you got from the leaders you talked to.

Step #6: Do More Homework

Before reaching the next major step—the formal proposal—you should find out if there are any constraints in developing the product. The constraints could be internal or external.

To determine if internal ones exist, ask youself:

- Is your company well enough funded?
- Do you have enough talented people?
- What about rights and royalties relating to the technology? The patent rights?
- How about your company's ability to produce, manufacture, distribute, and market the product?
- Will this product fit in with other company products—the current product mix?
- Who is going to manufacture your product? You or subcontractors? Are the right manufacturing facilities or subcontractors available?

To look for external constraints, ask:

- What's the competition like?
- What about the size of the market or client base?
- How are economic conditions?
- Are prospective suppliers available?
- How about the technological environment—the availability of raw goods or needed finished products?
- What about the availability of products associated with the use of the new one you propose to offer?
- Will plenty of spare parts be around?
- What about maintenance people?

You also have to look at another category of external factors: the legal and regulatory restrictions affecting the manufacture, sale, and use of the product.

This information is vital. Why? Because you have to answer questions from investors and perhaps your bosses.

Without the right background information, executives and investors won't believe in your seriousness. Your bosses might even question your competence in unrelated matters.

Step #7: Prepare a Prototype If You Haven't Done So Already

If you're really interested in your product, if you really believe in it by this time, then you should invest in developing a good operational prototype.

With something tangible, you can shortcut much investigation, as Wicket did in the case of Sherlock (the chemical leak-detector). You might save time and money. And your potential buyers can more accurately tell you how they'll respond to a product.

Step #8: Prepare a Market Engineering Proposal for Yourself, Your Bosses, and Prospective Investors

Your prototype is ready for a dog-and-pony show for your boss or investors. But you still must prepare a report, a corporate proposal.

It won't be a detailed market and business plan or final set of specs. Rather it will summarize all technical, financial, management, and marketing facts that your backers will need to make a "go-no go" decision. Keep the whole proposal shorter than 15 pages.

Here is how you should organize this market engineering proposal:

Definition of the Product
Use single sentences to describe:

- Its features and capabilities.
- Its variability. What about different editions or versions of the product?
- Potentials for enhancements or improvement.
- The expected life cycle. A software product's life cycle might be less than a year before a better rival overtakes your baby. The marketing life of equipment could vary. For a portable computer it might be less than a year. For a mainframe, it could last eight years, depending on sophistication and competing products.
- How long it will take to get the product to market. Some may require up to a year or two—which may be too late to capture a significant part of the market.
- Any benefits to be gained by using the product; for example, cost performance, reduction of problems, or reduction of security constraints.

Product Market Features
What are the product market features?

In other words, what are the markets by name and number (or quantity) and by size and segment? What are the niches? Will you be the main show, or a competitor?

The main show? Then your competitors might be on your tail in less than a year and pick off a piece of the prospective market pie, with lower-priced offerings—perhaps through reverse engineering. With rivals in mind, your report might say if you'll customize any, or all, of the product.

In this same category, you should estimate the costs of developing and marketing the product. And how about technology? What kinds will you need to build, use, test, and market your product?

Demands
What's driving this market? The original need or one for refinement?

Are you really coming up with a first-of-kind product in an area in which no one is demanding a better mouse trap?

Or is your mouse trap a catch-up—maybe an electronic version? Is it a smart enhancement of an existing technique that people have been looking for for a long, long time?

Here are other questions to answer about the buyers' needs, problems, and influences on prospective purchasers:

- What might have spurred interest? A government regulation, maybe? Perhaps part of the marketplace must buy certain products to meet government regulations.
- How much longer can your customers exist without your product?
- Are the current products satisfactory, or are they rife with problems?
- Will you have to convince the buyers of the need for a change—to your product?
- Do you think you can spark a clear demand as soon as buyers appreciate their needs? Is an industry already addressing such a need—and doing so in an organized manner, with statements from trade associations, users groups or others? Then this force can drive the marketplace.

Next you should summarize some of the classic kinds of marketing intelligence information. What about opportunities and threats? How about a changing government regulation, or the increased number of industrial accidents requiring better safety devices, or the existence of popular but unreliable rivals?

All these are opportunities for you.

Theats? They could be anything from major suppliers going bankrupt to government regulations that would work against the product.

Principal Competitors

Identify major rivals' sizes, their products, their product mixes, and their market shares for the products that are almost identical to yours. Also, note which rivals will be able to respond quickly with competing products—substitutes that can do the same job, perhaps more cheaply or faster. Maybe you'll sell a computer with a clock speed of 6 megahertz but your rival will fight back with a system running at 8 and selling for much less. In high tech, as the designers of the IBM PC AT discovered, that would be a plausible substitute project.

Of course, your proposal might note the differentials in capability, features, price, and benefits between your product and all competing products.

There are also the timing factors. Will the potentially competing product be on the market 6 to 20 months after yours, or at the same time, or before?

Barriers to Entry

You should list those situations or conditions that can prevent you from developing the product, from marketing and selling your product, and making a substantial income.

Do you have enough capital invested? The proper experience? A large enough laboratory? The right techniques or technology? How about the size of the market share you'll expect? Just what exactly are the risks?

How about government regulations and economic conditions in your industry or your clients' fields? Could they slow or even block your entry into the product market?

Bargaining Powers

Look at the bargaining powers of your buyers and your suppliers—see Chapter 4—and how they influence your product and the marketplace. The bargaining power of buyers includes whether or not they'll give you detailed specifications for the product and make you modify it. Or will they accept standard products?

And how about the suppliers? What about the reasonableness and stability of their prices? Will they deliver fast? Are they reliable? Will they have too much bargaining power?

Competitive Strategy for Gaining a Market Position

Take a look at your competitive strategy for gaining a market position share in order to make money. What positioning do you need? What's the forecast for the market for similar products?

Who has the shares, and how well are they doing? Here, you must consider "the key factors," as they're called—the main attributes of success. Are you looking for cost leadership in the field, for quality and service, or for delivery, or production, or materials?

How can you capture new business and increase profit margins? And how can you build them into your product and into your market plan?

Organizational Responses

How are you going to orchestrate product development and marketing? Other questions:

- What corporate resources do you need?
- When do you need them?
- What kind of people?
- What kind of space?
- What kind of equipment?
- What kind of talent do you need?
- How is it going to be managed?
- Who will be your product manager? You need one.
- Who will offer legal advice to help guard against conflicting trademarks or patents?
- Who will do the purchasing if you're going to buy supplies?
- Who will develop the sales force and the polished, professional-looking sales literature—the kind that most products manufacturers will distribute? Too many professional engineering services companies content themselves with skimpy or slovenly assembled documents that describe products and their uses and capabilities but don't really highlight the true advantages to buyers.

Without mapping out a good marketing engineering report that anticipates complications, you might risk severe financial setbacks. Consider the Poppett Company. It specialized in designing equipment that controlled industrial processes such as regulating the flow of gas through pipelines. Then Jay Poppett's brother-in-law asked him to design an energy monitoring and control system for a small, three-story building. Poppett jumped.

What a chance to break into this field and put out a product for buildings. So Poppett entered the energy control business—much to his regret.

He was behind schedule in the job for his brother-in-law but thought, "No problem. I'll get it better next time."

But he didn't. He failed in a bigger way, having unsuccessfully extrapolated from his experiences with his brother-in-law. He bid $15 million on a job on a 17-story building and expected a $1 million profit. Costs, however, skyrocketed to $23 million, and Poppett also muffed other major work.

Jay Poppett is now working for his brother-in-law. Lesson: Give the market engineering report the attention it deserves. Don't depend on earlier work for the necessary information and insights.

Step #9: Decide How You'll Create Your Success Model

Start thinking about how you'll organize your Success Model. I suggest seven parts—each of which I'll include in separate steps:

Step #10: Develop market engineering specs.

Step #11: Do your product engineering.

Step #12: Work out general marketing strategy.

Step #13: Consult with corporate officers and with investors.

Step #14: Do final engineering.

Step #15: Start up your marketing and sales promotion plans.

Step #16: Crank up your product test plan and respond accordingly.

Then you'll be ready for #17—putting your product on the market, and tracking its progress there.

Step #10: Market Engineering Specifications

The market engineering specification is, of course, a comprehensive market product plan for a specific product and a specific market.

Your earlier market engineering proposal will influence the market engineering specs and, in fact, might show up in them. The specs include three elements.

Element #1: A Basic Marketplace Definition

This definition is a true amplification of your market engineering proposal. Include such additional information as: (1) whether you've found more competition, (2) additional assessment of the technology, (3) new observations on product market mix or demands and needs, and (4) your basic overview of your product definition.

Element #2: Your Environmental and Organizational Response

Namely, how are you going to put this program across? What do you need within your own organization? Which people do you require to market and sell the product? Don't assign a petroleum engineer to market hospital computer systems. Believe it or not, that happened.

Also decide:

- When you'll need people in sales and marketing.
- What kind of management structure will oversee the sales and marketing.
- Who will be the product manager and the engineering manager.
- Whether you'll have a full-fledged research and development department.
- Who will handle administrative and legal details.
- What kind of buildings and other facilities you might need.
- What kind of training.
- Whether constraints in the organization would endanger a successful engineering and marketing program.

In other words, this section again requires you to develop the complete orchestration for engineering and marketing the product to ensure success.

Element #3: Financial Environment
This is just, in a sense, a strategic section. We won't get into pro forma statements yet; we're talking about requirements for marketing, for engineering, for all other resources—the rationale for capital requirements. This also includes your criteria for pricing, which you base on your general costs.

Step #11: Product Engineering

You're telling what the product will be like technically. So you offer two elements here: (1) product overview and (2) product specs.

Element #1: Product Overview
The marketing engineer or the marketer drafts the product overview based on the drives and the demands of the marketplace.
It includes:

- All of a product's features and configurations.
- Its strengths and weaknesses.
- Its life cycle—how long before another product will knock it out of the marketplace. In other words, this product overview covers all the features and the configurations.
- Who your suppliers are going to be.
- What kind of inventories you'll need.
- What kind of instruction manuals and other documentation you must provide.
- Details of manufacturing, including quality and cost control. You want to limit the number of parts without harming your price or performance, so your product will be cheaper to make. And probably your manufacturing experts[7] can think of many other details. Listen. You can

------

[7] Or a prospective subcontractor's.

reduce their headaches and yours if you consider all the details ahead of time.

- Information on reliability and maintainability. Is the product supposed to be extra-reliable? Is reliability the key factor for success? Even if it isn't, you'd better consult with the people in charge of servicing your miracle product. Find out (1) what's most likely to go wrong, (2) how you can design the product to reduce such breakdowns, and (3) how you can reduce the cost of repairs. If you're repairing a car, you normally don't have to tear the motor apart to replace a spark plug. You'd be surprised, though, at the number of high tech products that seem to be designed to rack up expensive repair bills.
- Facts on who will provide the service.

In other words, the product overview is a multidimensional profile of the product that's going to drive the next section, which is called product specification. And if your product is a software package? Then you'll need the software requirements, specifications, and maintenance documentation, which include performance characteristics.

Element #2: Product Specifications

Here you lay out all the detailed requirements of the product including:

- Its purpose, the functions it will perform.
- Performance specifications. How fast must it go? What kind of storage must it have? How high should it be? How's it to be used? What about environmental specs— operating temperatures or heating or air conditioning needs? Any software support services required? What hardware components do you need? Those are some of the issues here.

In other words, you have specs that reflect the conclusions expressed in your product engineering section. Product specifications might include your drawings, your specifications, your wiring diagrams. Creating software, obviously, you'd use a different style specs. But the idea would be the same—to describe the product accurately and thoroughly. Without good specs, you can't manufacture a marketable product.

Step #12: General Marketing Strategy

Your next step is your major marketing strategy. Answer these questions:

- How long will you need to get the product to the market and to make a profit?
- How are you going to position the product to win? What will you say is special about it? If nothing's special, you'd better go back to the drawing board.
- What do you do if your marketing plans go awry? Suppose the product is inferior or late, or what if a competitor comes barging in? You need a contingency plan based on the risks and the threats that you've identified earlier in your set of marketing and engineering specs.

Step #13: Consulting with Corporate Officers and Investors

In general, you're laying out your basic strategies and all your financial statements (your pro forma statements) including your costs, your budgets, your profits, and your taxes for all of your line items.

What to do next with this series of marketing and engineering specifications? Simple. Pass on the results of your research to corporate officers or your venture capital investors.

And use your research as a guideline for staffing after you get your money. Of course, the prime objective of this plan is to win your funds.

A second goal is the creation of a management control tool—a plan that tells you what you must do step by step, when you expect to do it, and how you'll spend your money.

Step #14: Final Engineering

Here, we're talking about product development.

Of course, at this stage you may get back into an operational prototype. Now, however, you're engaging in the very serious endeavor of building a product that you can make at a particular cost in accordance with the marketing and engineering specifications.

You might assign some engineers to plan the manufacturing process. Or they can help find companies that can put the product together for you after you have done the design.

Throughout this period of time, there is a continuing set of milestones for reviewing the progress of the engineering development that finally end up in the developmental test of the product.

Step #15: Marketing and Sales Promotion

In parallel with the engineering, you should start up your market and sales promotion programs.

That encompasses planning the test marketing of the product and of the sales promotion programs (including the preparation of a more detailed sales promotion program).

Here you lay out your objectives for the marketing and sales programs—your objectives and your goals. You go through the standard set of market program characteristics. Answer:

- What kind of communications media are you going to use? Will you run ads? If so, where? And what about possible press releases for trade journals or other publications? Also, will you communicate through trade shows, or only by taking the products on the road?
- Who is your audience?
- What messages will you disseminate?
- What kind of advertising are you going to use? What agencies?
- What trade shows are you going to attend to demonstrate your product?
- What kind of sales force do you need?
- How large will the sales force be, and where will it be distributed?
- What kind of distribution program will you have?
- Will you have your own distribution, or will you have to use an outside distributor?
- Will you look for value-added resellers who will take your

product and incorporate it into other systems, or other facilities?
- Are you going to have direct sales yourself with a sales force of so many people distributed throughout the country to push the product through the market segment you have selected?
- Are you going to look for distributors or dealers to do this?
- What about your costs of marketing and sales promotion?
- What about the scheduling of such activities?

Step #16: The Product Test Plan

Again, for the test plan, you've got to determine:
- Who your market is.
- What the audience is.
- What the method and timing for testing is.
- Selection of a test site (or sites).
- What kind of product demonstrations you are going to provide.
- The schedules of the demos.
- What kind of resources you need.
- What costs you are going to incur at what particular time.
- What the product budget is.
- What kind of feedback you want from the people who will test the product. Among other things, you'll obviously want their reactions to its characteristics and price.
- What kind of an adoption process are you going to expect? That is, how will the buying organizations and their employees incorporate your product into workday routines? What if you hope to sell computers to a company where everyone uses manual typewriters? How can you design the adoption process to disrupt the routine as little as possible? You might start by including the normal QWERTY keyboard. And the software commands might be identical to the way a manual typist would think and act when she's tapping away on her old Royal.

Now ask your clients for feedback—and respond accordingly.

Step #17: Get the Product on the Market—and See If Your People Are Moving It Well

Congratulations. You've developed a successful product and a successful marketing program.

Now your product is on sale.

But your market control program continues. You need to track the triumphs (or lack of them) of the marketing and sales program.

You'll deal with such issues as:

- Volume of sales.
- Share of the market for similar products.
- Market-expenses-to-sales ratios.
- Product profitability.
- The efficiency of the sales force.

Use these and other market and sales indices to track your progress and learn the extent of acceptance of your new product.

SUMMING UP

By encouraging innovation and following up on ideas methodically, services companies won't automatically produce winning products. But failure to do so will greatly increase the chances of Philistine-style failures.

Do's

    1. Check out product ideas: concepts, and prototypes with leaders of the client community. Those from your industry might either reject or steal them.

    2. Always verify the technical adaptability and compatibility of a new product. Unless you're IBM, for instance, don't offer computers with exotic, untested operating systems.

    3. Make sure that you back new products with training, distribution, telephone hotlines, and other services such as maintenance, warranties, and good documentation.

4. Let the marketplace—trends, demands, user needs—drive your product design.

5. Make sure your product is useful to a large enough market to justify your effort. If you beat swords into trophies, you can only market them to ex-soldiers.

Don'ts

1. Don't confuse wants with needs. Needs always exist. Wants are a variable. In other words, you basically satisfy needs in terms of what the client wants. Some may want a Rolls Royce to meet a need for transportation—but settle for a motorcycle.

2. Do not pursue product development without sufficient investment. New product development has inherent risks.

3. Don't confuse the technical success of a product with its market success. The drive to put the product on the market must coincide with marketplace trends and demands.

4. Do not wait until demands or trends reach a peak before introducing a new product—at least, not unless it has a spectacular benefit. Otherwise your market share and return on investment will be minimal. You'll have been just a follower of trends.

5. Don't expect overnight success. Making it requires at least testimonials from opinion makers.

AFTERWORD

A FEW FINAL THOUGHTS
(ON GLOBAL SNARK GAMES)

Too many Philistine Techs are out there. Whether it's marketing products or services, our high tech services industry has badly bungled in recent years.[1] Don't believe me? Then follow one of the rules of this book and compare us against the competition, the overseas kind in this case.

Having taken away millions of our factory jobs, foreign countries are confidently poised to attack laggards such as Philistine. "Asia," according to a recent *Washington Post* article, "is now turning to areas where American companies have remained dominant—innovation, product development and the service industries." Yes! Even the Star Wars program will apparently use generous helpings of Japanese technology. And already some U.S. programming jobs are going to China.

Thankfully, however, our problems are screamingly evident: misdirected salesmanship; haphazard marketing; and, above all, lack of competitiveness.

Zero defect marketing is not a panacea. What it can do, however, is at least reduce the chances of failure by encouraging high tech service companies to *serve*—through innovation and otherwise.

They have so much to offer. Both our government and our larger corporations need the expertise that only independent specialists can provide, ideally those with the competitive suc-

---

[1] Exceptions such as Electronic Data Systems are all too rare.

cess formula in their hearts and minds. Consider the trade deficit. How should we reduce it? By import quotas of the kind that helped bring on the Great Depression? By other nearsighted actions? By cutting back wages until workers earn no more than teenagers in Asian sweatshops?

No, the answer is *innovation in marketing and marketing of innovation*—the promotion of the economy and quality that technology can help bring about. Through services companies, Detroit can pick the brains of the specialists whom Ford or GM might not need constantly, but who can be useful in accomplishing specific tasks.

And the same would hold true in other corporate suites and in Washington. Private contractors can help turn our best thinkers loose on the great problems, while keeping down the size of the permanent government bureaucracy.

America's high tech services companies, then, mustn't wither away.

Rather they must learn to compete, not just against themselves but also against their foreign counterparts whose own genius may cost yet more American jobs and pride. Anyone for a good game of Snark.

Lee A. Friedman
Georgetown
Washington, D.C.
June 25, 1987

APPENDIX A

BID/NO-BID CHECKLIST

COMPANY NAME
COMPETITIVE MARKETING PROGRAM
DATE
LEAD/SOLICITATION TITLE: _____
RESPONSIBLE INDIVIDUAL: _____

BID/NO-BID CHECKLIST

Item	Hi Positive (Yes)	Med	Very Negative (No)
1. Is this a prime company target?	___	___	___
2. Have we been tracking?	___	___	___
3. Were we prepared for RFP?	___	___	___
4. Do we thoroughly understand the requirements?	___	___	___
5. Do we have necessary corporate experience?	___	___	___
6. Do we have necessary staff experience?	___	___	___
7. Do we have necessary management experience?	___	___	___
8. Are there areas where we have no experience?	___	___	___
9. If we bid will there be an impact on existing or planned contracts?	___	___	___
10. Can we overcome corporate weaknesses?	___	___	___

Item	Hi Positive (Yes)	Med	Very Negative (No)
11. Do we have qualified staff to write the proposal?	___	___	___
12. Do we have a qualified proposal manager?	___	___	___
13. Have we discussed our approach/ solution with key client decision-makers?	___	___	___
14. Did we discuss alternative approaches?	___	___	___
15. Have we prepared position papers on special technical areas?	___	___	___
16. Did we contribute to the SOW or specifications?	___	___	___
17. Did we review a draft of the SOW?	___	___	___
18. Does the contracting officer know us?	___	___	___
19. Do the technical people know us?	___	___	___
20. Does the client know our strengths?	___	___	___
21. Does the client know our weaknesses?	___	___	___
22. Do we know the client's preferences?	___	___	___
23. Do we know the client's buying reasons?	___	___	___
24. Do we know the client's concerns and biases?	___	___	___
25. Are we the client's favorite?	___	___	___
26. Do we know the competition?	___	___	___
27. Do we have a technical edge over the competition?	___	___	___
28. Do we have an experience edge over the competition?	___	___	___
29. Has company performed this type of work before?	___	___	___
30. Will this program maximize profits (e.g., no investment needed)?	___	___	___
31. Can we meet/control technical costs?	___	___	___
32. Do we have winning/competitive price?	___	___	___

Item	Hi Positive (Yes)	Med	Very Negative (No)
33. Do we have design/development tools?	___	___	___
34. Have we verified project's concepts, design, specifications?	___	___	___
35. Have we developed strawman proposal?			
Are there risks:			
36. Capital investment needed	___	___	___
37. Technical	___	___	___
38. Supplier	___	___	___
39. Management and Administration	___	___	___
How will we score on evaluation criteria:			
40. Requirements definition	___	___	___
41. Design/solution	___	___	___
42. Technical approach/methods	___	___	___
43. Management	___	___	___
44. Staffing	___	___	___
45. Corporate experience	___	___	___
46. Schedule	___	___	___
47. Price	___	___	___
48. Is evaluation technical : Cost ratio 50–50?	___	___	___
49. Does the client view us as a potential contender?	___	___	___
50. Does the client have a history of ECPs?	___	___	___
51. Do we have discriminators that ensure us winning?	___	___	___
1. Is the program funded? Are funds committed?	___		___
2. Do we know the budget?	___		___
3. Is there an incumbent?	___		___
4. Is this a recompete?	___		___
5. Are we the incumbent?	___		___
6. Have we worked for this client before?	___		___
7. Do we have copy of the current contract?	___		___

Item	Hi Positive (Yes)	Med	Very Negative (No)
8. Have we met with source selection people?	___		___
9. Has client met our proposed project mgr.?	___		___
10. Have we discussed the WBS with key decision-makers?	___		___
11. Have we discussed this opportunity with them?	___		___
12. Has all hardware been identified?	___		___
13. Has all software been identified?	___		___
14. Is a subcontractor necessary?	___		___
15. Have we signed an agreement with one?	___		___
16. Have we made joint marketing calls?	___		___
17. Does client have history of awarding to lowest bidder?	___		___
18. Will technical people "override" procurement (cost) people?	___		___
19. Do we have a pricing strategy?	___		___
20. Do we have a win strategy?	___		___
21. Do we know procurement processes?	___		___
22. Do we know procurement schedule?	___		___
23. Have we done strawman proposal?	___		___

For numeric indexing and to determine proposal probability:

High Positive Score = 10, 9, or 8
Medium Score   = 7, 6, 5, 4, or 3
Negative Score   = 2, 1, or 0
Probability Index   = $\dfrac{\text{Total of All Scores}}{\text{No. of Items} \times 10}$

Example: Assigning an average of 8 to each of the 73 items = 584

$$\frac{584}{730} = .80$$

APPENDIX B

AN EXAMPLE OF A RESPONSE MATRIX

SOLICITATION STATEMENT OF WORK WITH SPECIFICATIONS

"The contractor shall design, acquire and install a workstation-based office automation system. Workstations shall be interconnected via a star topology local area network(LAN), in accordance with IEEE standard 802.5. Each workstation shall be IBM compatible 80286 microprocessor-based, with a minimum of 640Kb memory, expandable to 1.6 MB. The workstation shall also include for disk storage a 20 MB removable hard disk (with approx. 65 millisecond access time), an AT-type keyboard, and EGA monitor (640×350 resolution), and the following commercially available software packages. . . ."

EXAMPLE
Compliance Matrix

Solicit. Reference Section	Requirement	Compliance Status	Proposal Reference Section
4.b.3	Design Workstation Office Automation System	yes	3.b,c.d; 4.c.2
	. . . Acquire	yes	5.d.2
	. . . Install	yes	5.d.6
	Workstation interconnect via star network	yes	2.c;3.b.1
	IEEE 802.5 standard	yes	3.b.2;4.b
	Workstation IBM compatible with 80286 processor	yes	4.c.2.1
	Memory at 640KB expandable to 1.6 MB	yes	4.c.2.2

EXAMPLE—*Concluded*

Solicit. Reference Section	Requirement	Compliance Status	Proposal Reference Section
	Disk storage of 20 MB removable HD (60 millisecond access)	no (tests show that 40 MB is needed for all software)	4.c.2.3 2.c.3
	AT Keyboard	yes	4.c.2.4
	EGA Monitor	yes	4.c.2.5
	Software (Pkg 1)	No (not compatible with operating system)	4.c.2.6.1

APPENDIX C

TYPICAL PROPOSAL OUTLINE

Note: As I note on page 203, your table of contents should go down at least four levels. This one—for the sake of conciseness—goes down only two or three.

TYPICAL PROPOSAL OUTLINE

COVER
(Title of project; client name/address; submittal date; your name/address)
A TITLE SHEET
(Repeat of cover plus proprietary/nondisclosure statement)
EXECUTIVE SUMMARY
COMPLIANCE OR RESPONSE MATRIX
TABLE OF CONTENTS

SECTION 1. INTRODUCTION
      a. *Submittal Declaration*
          (Cites solicitation number, title, date, and official amendments)
      b. *(Company Name) Overview*
          (Brief history of company, why qualified to do work, key commitments)
      c. *Scope of Proposal*
          (Section titles; contents of volumes)
SECTION 2. UNDERSTANDING OF REQUIREMENTS/PROBLEMS
      a. *Objectives/Goals of Program*
      b. *Analysis of Requirements/Problem*
      c. *Proposed Solution (or Support Services)*
      d. *Alternative Solutions Considered*

           e. *Benefits of Solution*
           f. *Risks*

SECTION 3. DESIGN (Of System, Study) (As per 2.a–c)
           a. *Design Concept, Rationale and Goals*
           b. *Proposed Architecture and Configuration*
           c. *Design Requirements/Specifications*
               (Performance, reliability, standards . . . by
               system, subsystem, and component)
           d. *The Design*
               (By function: operations, use, management,
               maintenance, person-system interface, other
               interfaces, and flows)

SECTION 4. TECHNICAL PLAN (To implement a design or a
system)
           a. *Objectives of Plan*
           b. *Technical Approach/Methodology Overview*
               (Tools, techniques, languages, standards . . . to
               be applied)
           c. *Methods*
               (As per statement of work for required design)
               1) Task 1 (objectives, methods, rationale,
                    deliverable products)
               2) Task *n*
               -OR-
               1) System Development
               2) Subsystem *A* Development (objectives, methods,
                    rationale, deliverable products)

SECTION 5. WORK PLAN
(Steps to accomplish Technical Plan)
           a. *General Requirements and Resources*
           b. *Work Teams and Assignments*
           c. *Work Controls and Reporting*
           d. *Work Procedures* (By Task)

SECTION 6. MANAGEMENT PLAN
           a. *Corporate Organization*
                (Structure/functions)
           b. *Proposed Project Organization*
                (Structure/function/interface)
           c. *Key Staff*
                (Functions/responsibilities)
           d. *Management Approach*
                (Controls, policies, procedures)

  e. *Quality Assurance/Controls*
  f. *Subcontract Management*
  g. *Contract Reporting*
  h. *Schedules*
   (By task, labor hours . . .)
  i. *Staffing Plan*
  j. *Deliverable Products Listed*
   (Including bill of materials)

SECTION 7. ASSUMPTIONS AND CONSTRAINTS
  a. *Assumptions* (responsibilities)
  b. *Constraints*
  c. *Risks and Resolutions*

SECTION 8. QUALIFICATIONS
  a. *(Company Name)*
   (Areas of specialty, client base, accomplishments, facilities)
  b. *Summaries of Relevant Contracts*
  c. *Staff Resumes*
  d. *Subcontractor/Supplier Qualifications*

[Note: A cost proposal normally shouldn't be part of a technical proposal. Rather it should be a separate volume. However, if your client wants, you can put the cost proposal in the technical volume. Be aware, however, of the advantages of separation. If you have a technical volume with a breakdown of costs and it falls into the hands of competitors, you're up the creek. So try to put the cost proposal in a different package—sealed.]

APPENDIX D

A PERT CHART
(To Help Schedule Various Tasks
That a Contract Calls for)

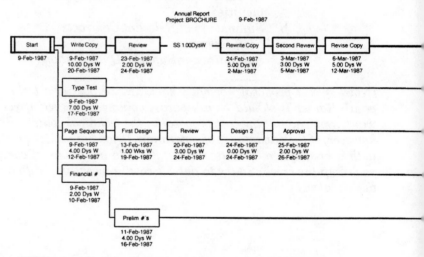

PERT courtesy of Harvard Total Project, Software Publishing Co.,
Mountain View, CA.

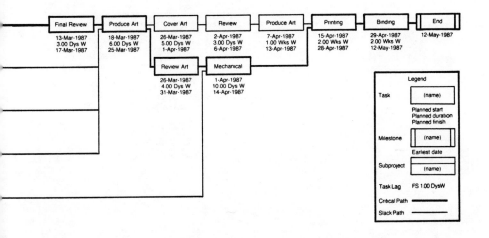

	Final Review	Produce Art	Cover Art	Review	Produce Art	Printing	Binding	End
	13-Mar-1987	18-Mar-1987	26-Mar-1987	2-Apr-1987	7-Apr-1987	15-Apr-1987	29-Apr-1987	12-May-1987
	3.00 Dys W	6.00 Dys W	5.00 Dys W	3.00 Dys W	1.00 Wks W	2.00 Wks W	2.00 Wks W	
	17-Mar-1987	25-Mar-1987	1-Apr-1987	6-Apr-1987	13-Apr-1987	28-Apr-1987	12-May-1987	

Review Art
26-Mar-1987
4.00 Dys W
31-Mar-1987

Mechanical
1-Apr-1987
10.00 Dys W
14-Apr-1987

Legend

Task	(name)

Planned start
Planned duration
Planned finish

Milestone	(name)

Earliest date

Subproject	(name)

Task Lag   FS 1.00 DysW

Critical Path ▬▬▬▬▬

Slack Path   ─────

INDEX